Changing the Faces of Mathematics

Perspectives on Indigenous People of North America

Series Editor

Walter G. Secada
University of Wisconsin—Madison
Madison, Wisconsin

Editor

Judith Elaine Hankes
University of Wisconsin—Oshkosh
Oshkosh, Wisconsin

Gerald R. Fast
University of Wisconsin—Oshkosh
Oshkosh, Wisconsin

National Council of Teachers of Mathematics
Reston, Virginia

Printed in the United States of America

Contents

Preface

In 1993, the National Council of Teachers of Mathematics created a task force on multiculturalism and gender in mathematics. That task force recommended the publication of a six-volume series, Changing the Faces of Mathematics, that would help make the slogans of "mathematics for all" and "everybody counts" realities. Each volume within this series has a different focus: The first published volume examined perspectives relating to Latinos; the next volume focused on perspectives on Asian Americans and Pacific Islanders; the third volume, perspectives on African Americans; the fourth volume, perspectives on multiculturalism and gender equity; and the fifth volume, perspectives on gender. This sixth volume is titled *Perspectives on Indigenous People of North America*.

The manuscripts selected for this sixth volume speak to many audiences: classroom teachers, administrators and principals, curriculum supervisors and program developers, and ethnomathematicians and researchers. The editors believe that investigators wishing to develop a deeper understanding of indigenous mathematics and pedagogy will find the volume a rich resource and that elementary, middle, and secondary teachers will be pleased with the many practical ideas shared by the contributing authors. Chapters 1 through 6 focus on the theoretical foundation that undergirds indigenous ethnomathematics; chapter 7 addresses the issue of portfolio assessment; chapters 8 through 17 include classroom lessons and activities that elementary and middle school teachers will find useful; and chapters 18 through 24 focus primarily on middle school and high school topics.

Of all cultural groups, the indigenous people of North America have the smallest percent of secondary and postsecondary students performing at the advanced level of mathematics. Consequently, they are underrepresented in careers that require higher level mathematics and science. One cause suggested for this inequity is, at least in part, the way that mathematics and science are conceptualized and taught in dominant-culture schools. However, another and more subtle contributing factor is reflected in the following response made by a sixty-one-year-old Oneida schoolteacher when questioned regarding her feelings about mathematics.

> I never was the greatest mathematician.... I never expected to be rich so that I would have to worry about how much money I might get.... The father works enough to keep bread on the table, we're fed, we're warm, all of that, but there's no looking down the road to say, "I'm going to have a bicycle by that time...." We are never going to be rich because we don't know how to, you know, how to hoard it. When someone needs something and we have it, we'll help. That's the way I am, but I really can't say me, because I'm not me. I am a part of a whole family.

The teacher, during an interview in her home, freely shared this response. The home attested to her sincerity. Although made comfortable with colorful throws and pillows, artificial flowers, and numerous family pictures, it spoke boldly against materialism and conspicuous consumption. The teacher's home and comments reflected her beliefs and values—beliefs and values, that influenced how she felt about mathematics.

One must not generalize that all indigenous people share the teacher's perspective. However, when reflecting on the mathematics instruction of Native students, a consideration of their cultural values is crucial. Social interactions that have a direct impact on mathematics-related attitudes might be as subtle as the preceding scenario. They might be obvious but informal, as when a grandfather helps a young gardener estimate the distance between rows of beans, or they might be obvious and formal, as when a classroom teacher administers a standardized test. Through such interactions, both spoken and unspoken cultural beliefs and values are transmitted to children. Understanding this process is intuitively and logically sensible, yet too often the implications of the process are ignored, and too often in mainstream school settings, culturally based beliefs and behaviors create misunderstanding and conflict. This volume has been written to serve as a resource for teachers who strive to teach mathematics to indigenous students in a culturally responsive manner.

Judith Hankes
Gerald R. Fast

Editors

Acknowledgments

This volume is the end result of the hard work of many contributors, including teachers and their students, mathematics educators, mathematicians, and researchers. The authors of the twenty-five chapters have unselfishly shared their expertise in mathematics pedagogy appropriate for indigenous peoples and have exhibited patience throughout the lengthy publication process of the volume.

We would like to extend a special thank-you to Walter G. Secada, Claudette Bradley, Jerry Lipka, and Jerilyn Grignon for reviewing manuscripts.

We would also like to express our appreciation to the editorial and production staff at the NCTM Headquarters Office for their help in converting the collection of manuscripts into a finished book.

Judith Hankes
Gerald R. Fast

Editors

Enhancing Mathematics Instruction for Indigenous American Students

1

Indigenous American students are underrepresented in careers requiring higher-level mathematics and science and, for that matter, in the requisite courses that prepare them for such careers (National Center for Education Statistics 1997). Evidence suggests that the cause of this inequity is at least in part the way in which mathematics and science are conceptualized and taught in American schools (American Indian Science and Engineering Society 1995; Bishop 1988; Grignon 1991; Nelson-Barber and Estrin 1995; Oakes 1990; Secada 1992). It is important to bear in mind that when we speak of indigenous American students, we are referring to students from a wide spectrum of tribes and settings, ranging geographically from the continental United States to Alaska and numerous islands in the Pacific. They may belong to intact homogeneous communities or be part of the diaspora of peoples forcibly removed from their homelands or pushed out by a variety of social pressures. They may live in urban, suburban, or rural environments. They may be strongly connected to tribal traditions and languages or less directly (but still significantly) affected by their ancestral communities' ways of thinking and communicating.

Despite these differences among indigenous groups, the experiences of indigenous students in the schools of the dominant U.S. society are similar in many respects. We hope that the reality of individual and group differences will not be obscured as we examine some of the similarities in how mathematics education may have failed indigenous students and how it may be improved for them. In education, consideration must always be given to the individual student and the social context within which he or she lives and learns. Our discussion is intended to assist in this process, not to replace it.

Current school reform initiatives have the potential to improve mathematics education for indigenous American students, *but only if specific attention is given to understanding their needs.* These reforms call for excellence and access for all students, but if they simply provide equal access to the existing system without an expectation that the system will be affected by new participants, they are likely to benefit neither indigenous students nor others from nondominant groups. Simply identifying standards acceptable to the mainstream and carefully teaching to those standards will not improve the learning of students who have been disfranchised in the past (Eisenhart, Finkel, and Marion 1996). Attention must be paid to making the content more meaningful by integrating more "context" into the curriculum.

Teaching Mathematics in Context

An important theme in efforts to reform mathematics education has been the need to connect school mathematics with children's own experience and

Elise Trumbull

Sharon Nelson-Barber

Jean Mitchell

INDIGENOUS
STUDENTS AND
SCHOOL REFORM

1

intuitive knowledge. However, when "context" is included in an attempt to "make the math meaningful," the particular context used is often simply an application of a skill or concept (e.g., inserting practice with operations into word problems, using given principles to construct a paper polyhedron or to design a swimming pool). These applications are often far from students' lived experiences, especially for indigenous students. Social or historical contexts or students' own cultural contexts are rarely brought to bear. If the goal of school mathematics should be, as Secada (1997) suggests, to "promote the development of those concepts that have the greatest validity and utility ... to students" (p. 8), we as educators need to find some way to incorporate social contexts in instruction.

Ideally, "learning occurs for the purpose of understanding and controlling not only school tasks but also the events in children's lives" (Palincsar 1989, p. 6). Attempts to solve the generic problem of providing meaningful context by incorporating only contexts from any given single cultural source will be insufficient. It is not enough for teachers and curriculum developers to become aware of such analyses of mathematics and sociocultural issues as those of Cocking and Mestre (1988) or Apple (1992). A need exists for more such analyses, especially from the cultural perspectives of the communities and groups affected, but not only from researchers and theorists. Teachers and students might also participate in this process to the benefit of all.

THE SHIFT TOWARD A CONSTRUCTIVIST VIEW OF LEARNING

School reforms over the past decade reflect a constructivist view of mathematics learning. In this view, individual students construct an understanding of mathematics concepts on the basis of their experiences within a community. At the same time, they participate in a social process of enculturation into school mathematics (Cobb 1994; Schoenfeld 1989; Bishop 1988). The student's personal experience base becomes a key to instruction and to understanding how students construe new information and experiences. A constructivist approach to education would regard students' culture-based experiences and ways of learning as resources for designing daily instruction that provides students with tools to address needs and solve problems of their own environment (see Aikenhead [1997]; Haidar [1997]).

Rethinking the Instructional Sequence

The traditional instructional sequence in mathematics has required students to master skills and facts first and then learn to apply them, but research suggests that students need to develop skills in the context of solving problems (Secada 1997; Carpenter 1985; National Council of Teachers of Mathematics [NCTM]1989; Lampert 1986). If so, the most common instructional sequence is not the most promising—for anyone. Once again, however, this generic concern has no generic solution, because mathematics instruction must connect with *specific* students' experiences. The activities from which concepts are to be developed must be meaningful to the students, not just the teacher or curriculum developer. Thus, constructivist teaching places demands on teachers to form essential bridges between students' lived experiences and the activities in their mathematics curriculum or to recast activities in contexts familiar to the students. Doing so requires teachers to have both a deep knowledge of mathematics concepts and more understanding of students' lives and cultures than traditional teaching has demanded.

Connecting to Students' Knowledge

Although the clear, logical implication of a constructivist view of learning is that school experiences and students' lived experiences should be connected, current interpretations of constructivism in mathematics curricular reforms often ignore the sociocultural dimension of students' knowledge. In fact, knowledge construction is not simply an individual act. It occurs within a social and a historical context (Eisenhart, Finkel, and Marion 1996). A "social constructivist" stance leads to concerns not only about curriculum content but also about organizational structures, approaches to interpersonal communication, instructional tools, and how all these reflect the world views and values of the school and of the students it is intended to serve.

For example, organizational structures both within and outside the classroom can either bring students and parents into the schooling process or discourage them from participating. If parents come from a culture that values consensus in making decisions that affect the whole group, they may be uncomfortable with a parent-teacher organization that requires them to vote as individuals without consulting others. They may prefer to hold a discussion until they can all come to agreement (Giancarlo Mercado, personal communication, 2 November 1996). Individual parent-teacher conferences may not be comfortable forums for some parents, who may not participate; a small-group format may work better (Trumbull et al. 1998, 2001). Schools that truly want to foster parents' involvement in their children's education need to examine the organizational structures and mechanisms that they put in place; sometimes these structures and mechanisms actually produce the opposite result from what they were designed to accomplish.

In response to the recommendations of current literature, a teacher may organize students into small cooperative learning groups. For students from some cultures, however, those groups will be much more functional if they are single-sex—the benefits of which the teacher (and current literature) may not be aware. With regard to interpersonal communication in the classroom, the teacher needs to know how a student's home culture expects children to participate in conversation and under what circumstances they are allowed to ask questions (and of whom). Classroom materials and teaching tools can, of course, encourage or discourage students' engagement. In notorious cases, teachers have used textbooks whose content requires an understanding of urban life in classrooms with rural students, without first building their background knowledge. In other instances, teachers have asked indigenous students to dissect in biology labs animals that are sacred to their groups. A more subtle example of a mismatch between a teaching strategy and the students it is intended to help is undue reliance on verbal explanations when visual representations would greatly enhance communication. This example certainly goes beyond cultural considerations; many students would benefit from a more multimodal approach to instruction.

Intuitive and Ethnomathematical Knowledge

It appears that we should reformulate our conception of mathematics teaching, viewing it as, in part, an exercise in helping children connect real-world experiences that require mathematical thinking with classroom mathematics. Through their real-world experiences, children develop intuitive mathematical knowledge—ways of thinking about problems involving spatial relationships, number, logical categories, and the like. Kieren (1992) has suggested that students' intuitive mathematical knowledge and ethnomathematics should be seen as the base on which technical symbolic knowledge and mathematical theory are built. We are using the term *ethnomathematics* to mean the forms of mathematics that are embedded in cultural activities in the workplace, the home, or in other community settings and are used by children or adults to carry out everyday tasks and solve everyday problems (Nunes 1992). (Examples are calculating the cost of goods, measuring cloth for making garments, and estimating the amount of food required to feed a group of people.) Ethnomathematical practices arise not only from the immediate needs of a group but also from its underlying values and beliefs about the world.

The ethnomathematical base of children's understanding should not be regarded as inferior or something to be discarded but rather as the heart of mathematical understanding. Children need to be able to connect formal symbolic representations with real objects, actions, and experiences, establishing correspondences between the real world and the representation of it (Kieren 1992). Figure 1.1 shows an instructional sequence reflecting this approach.

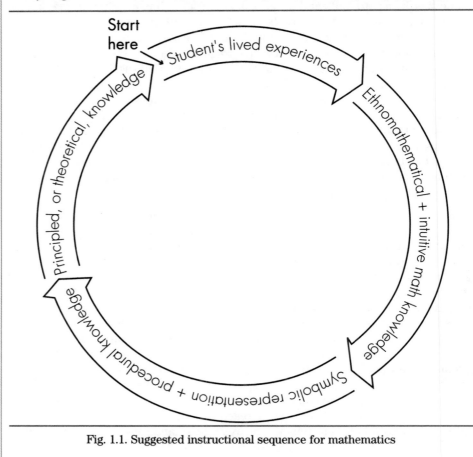

Note: Other relationships are likely among the elements in the circle. At times, mathematics learned in school may influence the interpretation of everyday experience and ethnomathematical thinking (though not as much as one might expect; see, e.g., Dowling, Paul, "The Contexualizing of Mathematics: Towards a Theoretical Map," in *Schools, Mathematics, and Work* [London: Falmer Press, 1991, pp. 93–120]). What is shown here is simply an instructional sequence.

Fig. 1.1. Suggested instructional sequence for mathematics

Room for many forms of mathematical knowledge

It is important to recognize that intuitive ways of solving problems or pragmatic routines that individuals devise to carry out computation or measurement are no less valuable than "school mathematics." Each serves its purpose. A role clearly exists for the "principled" mathematics that we tend to think of as the most abstract (involving as it does an understanding of mathematical theory), but it should not displace the intuitive and ethnomathematical bases that children develop through their own experience. Much of what gets characterized as "concrete" (such as the use of manipulatives) is actually quite abstract. (A block representing a fraction of a whole is, in itself, an abstraction from reality.) In our ideal vision of mathematics instruction, the following components would be included: (1) ethnomathematical knowledge and intuitive mathematical knowledge (i.e., mathematics based on experience), (2) symbolic representation and procedural knowledge, and (3) principled, or theoretical, knowledge. Instruction would begin with the identification of students' experiences that could form the basis for mathematical knowledge, and linkages would be made between the procedural and the theoretical realms. Instruction would not necessarily be linear; as suggested, it could move among the ethnomathematical and intuitive, the symbolic and procedural, and the theoretical.

Teachers have an essential role to play in incorporating appropriate ethnomathematical knowledge into their mathematics programs. For example, fifth-grade teachers in Chinle, Arizona, attempted to help their students move through the kind of instructional cycle shown in figure 1.1 (Koelsch, Estrin, and Farr 1995). They developed the Rug Task (see fig. 1.2) to show Navajo students how drawing a symmetrical design on a grid could be understood through patterns in woven rugs—an art form with which the students were intimately familiar. Many teachers enhanced the task by using it to review the concept of symmetry and to have students calculate area.

Recognizing students' knowledge

A third-grade teacher on the Pacific island of Rota told of taking her students on a field trip to a farm to learn about local agricultural practices. To her surprise, students who did not yet know about measuring angles were able to show her and the expert farmer the exact angle at which to plant on a hillside so that the rain would not wash away the seeds. She was able to take advantage of her students' knowledge in a subsequent lesson on angles and reminded the students over the course of instruction to think about what they already knew.

Unfortunately, indigenous students' intuitive and experience-based mathematical knowledge is not usually tapped in school. Mathematics as taught is not contextualized in a way that links it to indigenous students' experience and ways of thinking. Nonindigenous teachers may assume that indigenous students have an impoverished experience base because it does not match that of mainstream American children. Especially if they are teaching under the traditional mathematics education paradigm, teachers may not look actively for indigenous students' mathematical knowledge, nor are they likely to teach concepts in a context as advocated in the NCTM *Standards*. We will illustrate how teaching in context would be more compatible with indigenous students' ways of learning. However, even if a teacher is using a curriculum that has been formulated in accordance with the *Standards*, it is unlikely to provide a context meaningful to most indigenous students. Additional links to personal experience and meaning need to be made.

UNDERSTANDING CULTURAL ORIENTATIONS

Mathematics itself is a product of culture, and the mathematics taught in schools represents a particular cultural orientation. It also represents a philosophy about the nature of mathematics and how it should be taught that is "Western." (We use the quotation marks to indicate that the mathematics taught in schools is not strictly *Western*, although it is often characterized as such. In particular, influences from the Middle East and Africa need to be acknowledged.) For the most part, the prevailing approach to mathematics instruction is a rationalist one. As such, it excludes consideration of the historical, cultural, sociopolitical, or environmental aspects of mathematical and technological education (D'Ambrosio 1992). Indigenous parents and educators want students to master the mathematics concepts and skills taught in schools so that they can have occupational choices and can participate fully in the wider society. But

DESIGNING AND WEAVING A RUG

Mathematics Assessment Task

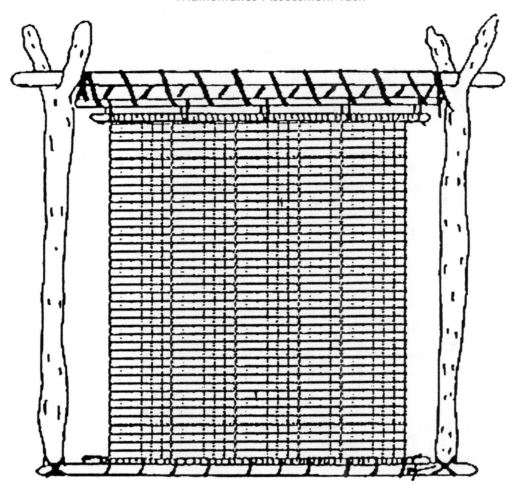

Instructions

You are spending the summer in the canyon with Grandmother. So that you can earn some money for new school clothes, Grandmother has offered to weave a rug for you to sell if you will create the design.

1. Design your rug in the grid on the following page [which shows a 5" × 7" grid with 1/4" squares for the students to use in their work].

2. Grandmother says it will take her five minutes to weave each row. How many minutes will it take her to weave your rug? Count the number of rows and multiply by 5. [Some teachers eliminate this statement to make the task more appropriately demanding.] Show your work below.

3. A tourist to the canyon came through on a Jeep tour and stopped to watch Grandmother weave. She was so impressed with Grandmother's weaving that she asked Grandmother to weave her a rug using her favorite pattern. [A simple pattern involving four squares is shown.] Use the pattern to design a rug. Use only this pattern in as many ways as you want.

4. Color in the tourist's rug using two or three colors.

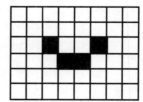

The Chinle rug task has three more parts that call on the student to make further calculations on the basis of the number of colors used and Grandmother's daily available time.

Fig. 1.2. The Rug Task

DESIGNING AND WEAVING A RUG—*Continued*

Name _____

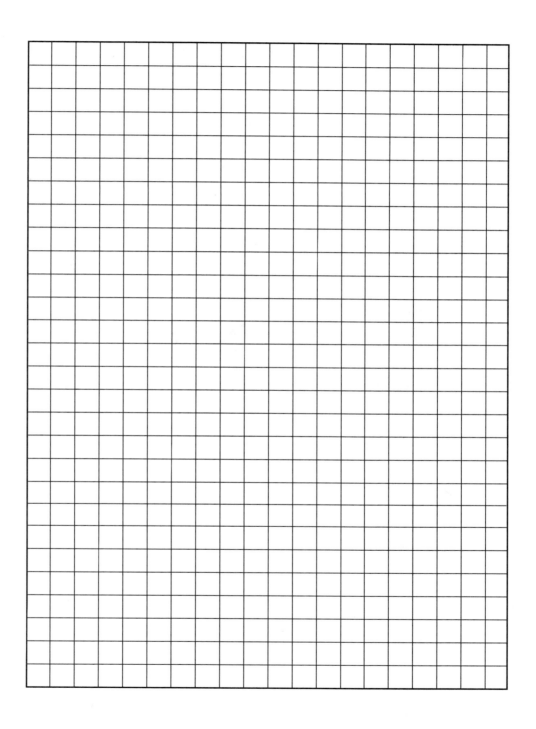

many also believe that the omission of ethical, social, and historical dimensions in framing curricula and standards in any subject is a serious oversight (Fox 1994). For example, those who want to consider the potential effects of a technological innovation on succeeding generations may not be comfortable with a context-free approach to innovation. When the important cultural values of indigenous students are not reflected in what they see of mathematics and science, the students may conclude that they must ignore some of their own values to participate in these domains. As a result, they may choose not to pursue these subjects at all.

Intertwined with the issue of the values in instruction are theories of what constitutes reality and the ways of knowing that reality. "Western" education makes particular assumptions about what is good, real, and true (cf. Scheurich and Young [1997]), often without making these assumptions explicit or recognizing that there may be other valid alternatives. We offer examples that we hope will help the reader think about how different ways of construing reality, different methods of learning about reality, and different values influence approaches to mathematics, teaching, and learning.

SIMILARITIES BETWEEN REFORM AND INDIGENOUS PEDAGOGIES

It is interesting that the approaches of indigenous peoples to teaching and learning coincide with some of the most highly touted elements of the research-based instruction called for by our nation's education reformers (see fig. 1.3). Concepts are taught in the contexts in which they will be needed; for example, numeration and quantification are taught in natural settings for practical purposes. Many indigenous groups emphasize learning by observation rather than by verbal rules or prescriptions. In addition, adults and older peers serve as models, guides, or facilitators rather than as direct instructors. Typically, children have considerable responsibility for their own learning, often working together in small groups to solve real-world problems or accomplish tasks. Children also have the latitude to choose when they will demonstrate their mastery of a particular task or skill, a situation that supports autonomy, self-evaluation, and perseverance until mastery is achieved (Au and Kawakami 1994; Lipka 1994; Swisher and Deyhle 1989).

Indigenous Pedagogy Reform Math and Science Pedagogy

Indigenous Pedagogy	Reform Math and Science Pedagogy
• Concepts are taught in meaningful contexts and serve authentic purposes.	• Concepts are taught in meaningful contexts, in more authentic ways.
• Adults serve as models and facilitators, guiding children to learn by observing and doing.	• Adults serve as models and facilitators; teachers are encouraged to go beyond strictly verbal methods of instruction.
• Children are encouraged to take responsibility for their own learning.	• Students are encouraged to take responsibility for their learning.
• Children are encouraged to evaluate their own learning.	• Students are encouraged to reflect on their own learning and engage in self-evaluation.
• Children are allowed choices about when and how to display learning (e.g., choices about being tested).	• New forms of assessment such as portfolios allow students more choice.

Fig. 1.3. Parallel features of indigenous and reform pedagogies

Elements in Potential Conflict

Other elements of reform pedagogy may not match the needs of indigenous students. Children used to learning through observation may not immediately take to a classroom in which the majority of learning and interaction is still based on verbal interchange. They may not readily translate a series of verbal instructions into a multistep activity (cf. Estrin and Nelson-Barber [1995] and Teacher Panel [1995] for discussions of the limitations of new performance assessments for indigenous students). They may also be less amenable to approaches calling for trial and error, which many discovery or inquiry methods advocated by mathematics reform efforts often use. Students may prefer to observe until they feel ready to try something themselves. Finally, children who have been taught to respect elders' knowledge without questioning may not be comfortable questioning their teachers as part of the process of developing critical-thinking skills. None of this should be taken to mean that mainstream approaches cannot be successful with these students. It does, however, imply that additional thought must be given to ways of introducing activities and motivating students or to adapting activities that have been identified as exemplary for mainstream students.

Even when teachers are contextualizing their mathematics teaching within culturally appropriate and personally relevant activities, students may not automatically make connections between classroom representations and "personal models of reality" (Kieren 1992). Therefore, some activities must be developed specifically for the purpose of making the connections explicit in students' minds. But if teachers are going to help students make these connections, they need to (*a*) recognize that students have their own ways of mathematizing the world and (*b*) discover ways to bring these models to the surface for discussion.

Direct Questioning

Before reviewing some more culture-based examples that we hope will bring life to the instructional model we propose, we must acknowledge that "getting into students' heads" to help them make their intuitive and ethnomathematics knowledge explicit is a challenge. The processes associated with these kinds of knowledge operate somewhat automatically in natural settings, and they may be difficult to bring to consciousness for analysis. One strategy is to ask students why and how they solved problems as they did, a technique called for by the Mathematics as Communication Standard (NCTM 1989). Such questions can lead to insights both about students' developmental levels in mathematics and about their specific ethnomathematical knowledge. However, this strategy is not without its limitations, for at least two reasons. First, sometimes students cannot—or will not—respond to such explicit probes. If such a strategy is to be productive, students first need to know that it is all right (and positively valued) to have alternative solution paths to a problem and that there is not just one correct or expected answer to such questions. Teachers and students alike need to understand that different experiences grounded in different cultural realities can lead to the same specific mathematics concepts—that there are "many roads up the mountain" (Mitchell 1996).

The second potential problem with a strategy of directly asking students how they solve problems has to do with language and with culturally based communication styles. Students may misinterpret a teacher's questions or even the reasons for them, or the teacher may misinterpret the student's response (cf. Eriks-

TAPPING
STUDENTS'
THINKING

Brophy and Crago [1993] and Philips [1983]). Students do not automatically know what questions mean, and questions often mean different things in the context of mathematics than they do in the context of natural language. What is considered an appropriate response to the question "Why?" or "How?" can be different in the context of a mathematics activity from what might be appropriate in some other setting. A simple question like "Why did you solve the problem that way?" does not call for a personal defense but a mathematical rationale—the kind of response that students learn how to craft over the course of many interchanges with their teachers.

How to help students engage in mathematical discourse is a problem for teachers even when they share both language and cultural background with their students (Pimm 1987; Laborde 1990). It is further complicated if, for instance, questioning a person about the reasons for his or her actions is not a normal part of that person's culture or if children are not used to being questioned by adults. For these reasons, teachers need to be particularly sensitive to their students' responses to being questioned, and they may even need explicitly to model and teach the kind of question-and-answer cycles that are appropriate for talking about mathematics. They also can develop more indirect ways of getting at students' thinking, such as eavesdropping on students' discussions with one another during group tasks. And at times they may need to tap resources outside their classrooms—in the students' communities or possibly the research literature—for insights into their students' thinking.

The Role of Relationships

Methods and curriculum are not the only factors in the successful mathematics instruction of indigenous American students. The relationships between teacher and student and among students are primary in any kind of school learning and are perceived by many to be of special significance for minority students (Nelson-Barber and Mitchell 1992). When students are given choices and their own experiences and modes of learning are incorporated into the classroom, students can become "equal partners in their own learning" (Bartolomé 1994, p. 186). In whatever ways teachers can accomplish this kind of partnership, they need to do so. Without it, no amount of specialized pedagogy is likely to result in students' success. We need to remember also that peers play a very important part in learning in most indigenous communities and that classroom structures that support peer-to-peer learning, such as cooperative learning, need to be used.

EXAMPLES OF ETHNO-MATHEMATICS IN INDIGENOUS CULTURES

Cross-cultural studies suggest that all societies are concerned with mathematical domains—number and quantification, space and time (and their measurement), and logical relationships—and have devised systems for organizing these elements. Such systems may look quite different from culture to culture (Ascher 1991; Bishop 1991). "The differences, however, are not [in] the ability to think abstractly or logically. They are in the subjects of thought, the cultural premises, and what situations call forth which thought processes"(Ascher 1991, p. 190). Ethnomathematics that is not recorded or formalized may be used by groups in a way that allows it to be passed on beyond the context of its immediate usefulness. Such systems will have arisen in very different settings associated with very different physical demands and world views and will naturally be appropriate to the purposes of the communities that devise them. Numerous examples of ethnomathematics have been documented throughout the world in such applications as making garments, doing construction work, packing food, and cooking, among other human endeavors (for ways to use such ethnomathematical examples in the classroom, see Ascher [1991]; Harris [1991]; Lave [1988]; Lipka [1994]; Nunes [1992]; Gerdes [1988a, 1988b, 1988c], Moore [1988a, 1988b]; Rauff [1996]; Zaslavsky [1973, 1990]).

The Treatment of Space and Time

The ethnomathematics and related world views of indigenous communities often differ significantly from those of "Western" culture in ways that affect how indigenous students relate to mathematics instruction. For example, the Inuit (from what is now Alaska) have traditionally treated time in a different way from Western culture. They do not consider it as an entity in itself but as part of a space-time dimension (Ascher 1991), reflecting a focus on interrelated processes rather than on isolated elements. The same is true for the Navajo. Because of the importance of motion and process, static measures are not valued in the same way that they are in "Western" mathematics. Moreover, the Inuit have not been concerned with organizing past events in chronological sequence: what happened in the past is regarded as an attribute of what is now.

The specific description of spatial location is extremely important for the Inuit, who may travel several hundred miles in an icy, snowy landscape that is continually changing because of the weather. It is more useful to them to take careful note of significant, unchanging aspects of the landscape (a hill, a ridge along a river, the configuration of a coastline) than to establish measures of distances in standardized units. In fact, the flexibility to take multiple perspectives, rather than the ability to use specific units of measurement, is valued (Ascher 1991). The language of the Inuit has the capacity to make descriptions of location extremely explicit. "Localizers" (adverbial suffixes) are added to verbs not only to indicate the specific orientations of an action (e.g., *in*, *up*, or *over there*) but also to show from whose perspective (e.g., that of the speaker, the addressee, or a third person) the actions and orientations are perceived. Inuit mathematics reflects these values and constraints.

Number Systems

Cultures have different ways of constructing number systems, and all use some form of visual aid, whether it be sticks, other kinds of counters, knotted string, or written symbols (Hinton 1994). A culture's words for numbers often reflect its visual counting system. According to Hinton (1994), some among the various California Indian groups used a decimal system (the base-ten system); others used a quinary, or base-five, system. Still others used a base-four system. People speaking Wintu or some of the Pomo languages used a vigesimal, or base-twenty, system or one that was at least partly vigesimal. Most of the Pomo languages used the word for "stick" to refer to multiples of 20 (Hinton 1994). The Iñupiaq and the Yup'ik Eskimos of Alaska also traditionally used a base-twenty system with a subbase of five. Virtually all number systems studied used the body—fingers, toes, spaces between the fingers and toes (apparently the origin of base-four systems), hands, feet, shoulders, arms, and the like—for points of reference in counting. Number systems may mix different bases because of intercultural encounters. For example, in the Yup'ik language, the counting system is "pure" between 1 and 400 but thereafter is a combination of base-twenty and base-ten systems, accommodating the influence of the Russian system introduced in the 1800s.

DEVELOPING INSTRUCTION AROUND ETHNO-MATHEMATICS

Examples from Alaska

An Iñupiaq perspective

The potential for using these indigenous systems for significant mathematical learning in school has been demonstrated by Clark Bartley, a junior high school mathematics teacher, and his Iñupiaq students in Alaska. This group explored arithmetic developed from the traditional Iñupiaq counting system, which is a base-twenty system with a subbase of five (Bingham 1997). The system Bartley and his students developed is now being used in a number of classrooms, whether Iñupiaq students are in the class or not. Perhaps in part because the system is new and the students are doing "real" mathematics and not simply trying to reproduce the knowledge of their teacher, they become very engaged and begin to take the initiative in developing their mathematical thinking. Anchorage teacher Paul Ruston says, "What I've noticed across the elementary schools is they'll spend forty-five minutes discovering new things about the system in class" (Bingham 1997). Bartley and his students are pursuing the cross-cultural connections of the system as well, communicating with people in Greenland, Baffin Island, and even Russia to find out if they have similar systems.

A clear benefit accrues to Iñupiaq students in the motivation and pride that come from working in a context that is meaningful to them. The students' development of an understanding of mathematical notation and algorithms and their connections to operations and concepts related to place value and number bases, although important, is only one aspect of the value of such an experience. Students also experience mathematical thinking firsthand; they have the genuine experience of doing mathematics and being mathematicians in the sense advocated by Brown, Collins, and Duguid (1989), Lampert (1990), Lave (1988), and Schoenfeld (1988).

A Yup'ik perspective

Another promising effort that uses ethnomathematics as the basis for a mathematics curriculum is ongoing in the area near Fairbanks, Alaska. A group of fourteen Yup'ik teachers, known as the "Ciulistet" (leaders) has been working closely with Yup'ik elders to develop a Yup'ik mathematics curriculum for use in the public schools (Lipka with Mohatt and the Ciulistet Group 1998). They have begun in part with the Yup'ik counting system, which uses the human body as a frame of reference with top-to-bottom and left-to-right axes for orientation. They note that if a drawing of the body with arms and legs outstretched is placed within a circle, other concepts, such as orientation and directionality; part-to-whole relationships (fractions); circles, arcs, and degrees; coordinate geometry; and symmetry, are implied (Lipka 1994). In the Yup'ik way of thinking, the body is a sensing instrument, giving a person a sense of time, space, and place (Kawagley 1995). Yup'ik ethnomathematics and ethnoscience begin with the body as an instrument and source for numeration, measurement, orientation, geometry, and symmetry (Lipka 1998).

Examples from Arizona

A Navajo perspective

Teachers in the Chinle Public Schools in Arizona, where 98 percent of the student body is Navajo, have created district standards for learning. One of these standards has to do with "Environmental and Cultural Awareness and

Responsibility" (Koelsch and Trumbull 1996; Koelsch, Estrin, and Farr 1995). This standard is invoked in numerous activities that call for and develop mathematics and science concepts and skills. As a public school district, Chinle must address the standards developed by the state of Arizona, as well, and show that its students are meeting them. Chinle educators have examined how state standards can best be addressed in their own setting while ensuring that Navajo values are respected in the curriculum.

WestEd, a regional research and development laboratory, has worked with teachers in Chinle for four years to develop a culturally-responsive assessment system. This system is aligned and integrated with the curriculum and culturally responsive instructional approaches, and it conforms to both the statewide and district culture-based standards. Assessments may take the form of a relatively short task (e.g., reading and responding to a piece of literature) or an extended project (e.g., learning about landfills), but they always maintain a strong link to the standards (including the culture-based ones), to the curriculum, and to formats and processes used in instruction.

The Mural Task was developed by the Chinle teachers for students in grades 6–8. It has been designed to be used as an assessment with criteria set for what elements must be included and how students will be judged on their performances. However, it is also a learning activity. The task overview for the teacher reads as follows:

> For this task, each student will have the opportunity to draw a design for a mural to be painted on the exterior of Building A. The design must include a geometric, Southwestern border framing a Navajo scene. The scene should reflect pride and respect for life on the Navajo Nation. The finished mural will be 80 inches by 160 inches. The student will develop a proposal for the design that includes a description of its cultural value, [along with] an estimate of cost and production time. The project will be part of mathematics, art, culture, and language arts classes and will be included in students' portfolios.

A team of teachers and students choose from among the students' designs two murals to be painted on the building. The prompt of the task guides students through a series of nine steps to complete a plan. One of the first steps is to conduct interviews with local people to get ideas about the subject matter for their mural. Students are urged to consider the possible roles of landmarks, people, cultural activities, livestock, plants, and legends in their murals. One can see how this kind of task is culture linked and contextualizes a number of academic goals. It is not based on a naturally encountered example of ethnomathematics, but it does allow students to use their own cultural knowledge.

Another task developed in Chinle, the Heritage Task, draws on cultural knowledge, has mathematical elements, and is itself a culture-based activity. Students chart their family lineage, using symbols and a hierarchical scheme that shows generations of family members and their clan membership. Such an activity can be a bridge to school-based mathematics concepts without itself having to be mathematized in ways that might be considered inauthentic or artificial. Teachers cannot assume, however, that students will automatically engage in the desired mathematical thinking without explicit instruction. That is, they may not see the links between the implicit mathematics in a naturally occurring activity and the explicit concepts that their teacher is introducing as "school mathematics."

Making the form of assessment culturally responsive

The assessment reform process in Chinle has included attention not only to content but also to forms of assessment and processes that are culturally responsive to the Navajo students and community. It is evident from the examples provided here how a local context has been used to situate school mathematics content and make it more meaningful or accessible to students. The district has also chosen forms of assessment that it believes make cultural sense as well: portfolios, projects, and relatively open-ended performance tasks. Through portfolios, selected work from students can be organized chronologically and according to the content standards (in whatever subject) that it demonstrates. Commentaries by teachers and students further clarify the meaning of a portfolio entry—why it was chosen, what it shows about students' learning, and under what circumstances it was completed. It is a highly contextualized collection of students' work. Products arising from students' projects (e.g., a piece of writing, a videotape, a three-dimensional model, a drawing) are also closely linked to a larger context; they are not decontextualized samples of knowledge or skill. The portfolio process makes more sense for these students than a series of formal, standardized tests administered on a pre-arranged schedule.

Assessment tasks take the form of miniprojects in that they require the integration of many skills in a meaningful activity. These forms of assessment, in contrast with multiple-choice or short-answer tests, have been recommended by American Indian educators as much more appropriate for finding out what their students know and can do (Estrin and Nelson-Barber 1995). Not incidentally, these forms of assessment also permit flexibility in administration. Performances are not timed, and they do not need to occur on a specific date but can be scheduled as they make sense. Students have some choice, often about the topics for their projects, about how they will demonstrate learning, and about when they will do so—an important feature, according to American Indian teachers (Estrin and Nelson-Barber 1995; Koelsch, Estrin, and Farr 1995; Teacher Panel 1994).

SUCCESSFUL PEDAGOGY FOR INDIGENOUS AMERICAN STUDENTS: SUMMARY OBSERVATIONS

In addition to the pedagogical skills and understandings detailed in NCTM's *Professional Standards for Teaching Mathematics* (1991), teachers need to have concrete knowledge of the kinds of mathematical understandings that indigenous students bring to school from their experience outside school. Only then can they begin to determine how to support the further development of students' understanding and make connections between students' knowledge and what might be called the mathematics "world heritage." Likewise, it is important to find culturally relevant ways of introducing new concepts and skills from this world heritage so that students will incorporate them into their own mathematics. Teachers must consider students' cultures to decide which ethnomathematical activities offer natural opportunities for mathematization so that students can develop these processes in a context that is meaningful to them.

Teachers need to understand more than just students' cultures. We agree with Bishop (1991) that teachers need to know about the values inherent in the subject they are teaching (or the values inherent in the particular perspective they are taking on the subject) and about the cultural history of that subject. They need to consider their own values and those of their students' cultures in relation to how and what they are teaching. These elements must be made explicit or conscious if mathematics pedagogy is to be culturally responsive.

But although—or perhaps because—teachers' understanding is ultimately the key to successful mathematics teaching and learning for indigenous students, teachers must not be left only to their own resources. Teachers cannot be expected to develop the requisite knowledge or understanding in a vacuum simply because they are told they "should" have it. The mathematical education most teachers have received did not include an examination of values and perspectives; seldom will it even have included any history of mathematics from which they might develop an appreciation of mathematics as a product of culture or any specific knowledge that they could connect with indigenous ethnomathematics in productive ways. Nor does a teacher's education or personal experience ordinarily include any knowledge of the kinds of cultural experiences or world views that indigenous students have as the core of their mathematical understanding (Kieren 1992). Finally, we have little reason to believe that all the relevant knowledge is currently part of the general knowledge base of the profession. Thus, both research and teacher education (preservice and inservice) have essential roles to play in developing an improved mathematics pedagogy for indigenous American students. In the meantime, teachers can continue to learn from their students and the communities from which their students come.

Both the move toward a more constructivist view of learning and the implementation of educational approaches embraced by mainstream reforms, such as thematic and project-based instruction, signal a genuine opportunity for improving mathematics instruction for indigenous American students. The incorporation of culture-based perspectives in the daily life of the classroom—a logical outgrowth of the constructivist view—can be accomplished when teachers develop a knowledge of their own and their students' approaches to mathematics (and ways of knowing in general). Mathematics, like all school subjects, needs to be seen by teachers and students as the set of cultural assumptions and activities it is.

When mathematics is taught and learned in defined cultural contexts, students have a greater opportunity to relate to it and find it meaningful. We believe that by drawing on students' own experiences that have elements of mathematics in them (experiences that are more and less mathematized) and by linking classroom learning to community-based practices, it is possible to foster a climate in which indigenous students will be less likely to reject mathematics. We believe that beginning with students' own experiences and moving toward formal classroom symbolization and procedures in mathematics (and then stepping back again to experiential context) is a more promising instructional sequence than the typical one is, which begins with the abstract and then moves on to applications.

REFERENCES

Aikenhead, Glen S. "Toward a First Nations Cross-Cultural Science and Technology Curriculum." *Science Education* 81 (1997): 217–38.

American Indian Science and Engineering Society. *Educating American Indian/Alaska Native Elementary and Secondary Students: Guidelines for Mathematics, Science and Technology Programs.* Boulder, Colo.: American Indian Science and Engineering Society, 1995.

Apple, Michael W. "Do the Standards Go Far Enough? Power, Policy, and Practice in Mathematics Education." *Journal for Research in Mathematics Education* 23 (1992): 412–43.

Ascher, Marcia. *Ethnomathematics: A Multicultural View of Mathematical Ideas.* Pacific Grove, Calif.: Brooks/Cole Publishing Co., 1991.

Au, Kathryn H., and Alice Kawakami. "Cultural Congruence in Instruction." In *Teaching Diverse Populations: Formulating a Knowledge Base*, edited by Etta R. Hollins, Joyce E. King, and Warren C. Hayman, pp. 5–23. New York: State University of New York, 1994.

Bartolomé, Lilia. "Beyond the Methods Fetish: Toward a Humanizing Pedagogy." *Harvard Educational Review* 64, no. 2 (1994): 173–94.

Bingham, Charles. "Teacher Spreads Word of Iñupiaq Math System." *The Arctic Sounder*, 1 May 1997, pp. 5, 8.

Bishop, Alan J. *Mathematical Enculturation: A Cultural Perspective on Mathematics Education.* Boston: Kluwer Academic Publishers, 1988.

———. "Mathematics Education in Its Cultural Context." In *Schools, Mathematics, and Work*, edited by Mary Harris, pp. 29–40. London: Falmer Press, 1991.

Brown, John Seeley, Allan Collins, and Paul Duguid. "Situated Cognition and the Culture of Learning." *Educational Researcher* 18, no. 1 (1989): 32–42.

Carpenter, Thomas P. *Learning to Add and Subtract: An Exercise in Problem Solving.* Madison, Wis.: University of Wisconsin, Wisconsin Center for Education Research, 1984.

Cobb, Paul. "Constructivism in Mathematics and Science Education." *Educational Researcher* 23, no. 7 (1994): 4.

Cocking, Rodney R., and José P. Mestre. "Considerations of Language Mediators of Mathematics Learning." In *Linguistic and Cultural Issues in Learning Mathematics*, edited by Rodney R. Cocking and José P. Mestre, pp. 3–16. Hillsdale N.J.: Lawrence Erlbaum Associates, 1988.

D'Ambrosio, Ubiratan. "What Does It Mean to Be Modern in Mathematics Education?" In *Developments in School Mathematics Education around the World: Proceedings of UCSMP International Conference on Mathematics Education*, October 30–November 1, 1991, vol. 3, edited by Izzak Wirzup and Robert Streit, pp. 155–62. Reston, Va.: National Council of Teachers of Mathematics, 1992.

Eisenhart, Margaret, Elizabeth Finkel, and Scott F. Marion. "Creating the Conditions for Scientific Literacy: A Re-examination." *American Educational Research Journal* 33, no. 2 (summer 1996): 261–95.

Eriks-Brophy, Alice, and Martha B. Crago. "Transforming Classroom Discourse: Forms of Evaluation in Inuit IR and IRe routines." Paper presented at the American Educational Research Association annual meeting, Atlanta, Ga., 12–16 April 1993.

Estrin, Elise Trumbull, and Sharon Nelson-Barber. *Issues in Cross-Cultural Assessment: American Indian and Alaska Native Students.* Knowledge Brief no. 12. San Francisco, Calif.: Far West Laboratory for Educational Research and Development, 1995.

Fox, Sandra. "National Indian Education Goals Panel." Paper presented at the annual meeting of the National Indian Education Association, Saint Paul, Minn., April 1994.

Gerdes, Paulus. "On Culture, Geometrical Thinking and Mathematical Education." *Educational Studies in Mathematics* 19, no. 2 (1988a): 137–62.

———. "On Mathematical Elements in the Tchokwe 'Sona' Tradition." *For the Learning of Mathematics* 10, no. 1 (1988b): 31–34.

———. "On Possible Uses of Traditional Angolan Sand Drawings in the Mathematics Classroom." *Educational Studies in Mathematics* 19, no. 1 (1988c): 3–22.

Grignon, Jacqueline. "People of Color: An American Indian Perspective." Paper presented at the annual meeting of the American Educational Research Association, Chicago, Ill., April 1991.

Haidar, Abdullateef H. "Western Science and Technology Education and the Arab World." In *Effects of Traditional Cosmology on Science Education*, edited by Masakata Ogawa, pp. 6–14. Mito, Japan: Ministry of Education, Culture, and Sports, Ibaraki University, 1997.

Harris, Mary, ed. *Schools, Mathematics and Work.* London: Falmer Press, 1991.

Hinton, Leanne. *Flutes of Fire: Essays on California Indian Languages.* Berkeley, Calif.: Heyday Books, 1994.

Kawagley, Oscar. *A Yup'ik Worldview.* Prospect Heights, Ill.: Waveland Press, 1995.

Kieren, Thomas E. "Rational and Fractional Numbers as Mathematical and Personal Knowledge: Implications for Curriculum and Instruction." In *Analysis of Arithmetic for Mathematics Teaching*, edited by Gaea Leinhardt, Ralph Putnam, and Rosemary A. Hattrup, pp. 323–71. Hillsdale, N.J.: Lawrence Erlbaum Associates, 1992.

Koelsch, Nanette, and Elise Trumbull Estrin, and Beverly Farr. *Guide to Developing Equitable Performance Assessments.* San Francisco: WestEd, 1995.

Koelsch, Nanette, and Elise Trumbull. "Portfolios: Bridging Cultural and Linguistic Worlds." In *Writing Portfolios in the Classroom: Policy and Practice, Promise and Peril*, edited by Robert C. Calfee and Pam Perfumo, pp. 261–84. Mahwah, N.J.: Lawrence Erlbaum Associates, 1996.

Laborde, Colette. "Language and Mathematics." In *Mathematics and Cognition: A Research Synthesis by the International Group for the Psychology of Mathematics Education*, edited by Perla Nesherand and Jeremy Kilpatrick, pp. 53–69. Cambridge, U.K.: Cambridge University Press, 1990.

Lampert, Magdalene. "Knowing, Doing, and Teaching Multiplication." Occasional paper no. 97. East Lansing, Mich.: Michigan State University Institute of Research on Teaching, 1986.

———. "When the Problem Is Not the Question and the Solution Is Not the Answer: Mathematical Knowing and Teaching." *American Educational Research Journal* 27, no. 1 (1990): 29–63.

Lave, Jean. *Cognition in Practice*. Cambridge, U.K.: Cambridge University Press, 1988.

Lipka, Jerry. "Culturally Negotiated Schooling: Toward a Yup'ik Mathematics." *Journal of American Indian Education* 33, no. 3 (1994): 14–30.

Lipka, Jerry, with Gerald V. Mohatt and the Ciulistet Group. "Expanding Curricular and Pedagogical Possibilities: Yup'ik-Based Mathematics, Science and Literacy." In *Transforming the Culture of Schools: Yup'ik Eskimo Examples*, edited by Jerry Lipka with Gerald V. Mohatt and the Ciulistet Group, pp. 131–81. Mahwah, N.J.: Lawrence Erlbaum Associates, 1998.

Mitchell, Jean. "Improving Concepts of Mathematics Education." Unpublished manuscript.

Moore, Charles G. "The Implication of String Figures for American Indian Mathematics Education." *Journal of American Indian Education* 28, no. 1 (1988a), 16–26.

———. *The Outdoor World Science and Mathematics Project*. Flagstaff, Ariz.: Science Learning Center, Northern Arizona University, 1988b.

National Center for Education Statistics (NCES). *Findings from the Condition of Education 1996: Minorities in Higher Education*. NCES 97–372. Washington, D.C.: NCES, 1997.

National Council of Teachers of Mathematics (NCTM). *Curriculum and Evaluation Standards for School Mathematics*. Reston, Va.: NCTM, 1989.

———. *Professional Standards for Teaching Mathematics*. Reston, Va.: NCTM, 1991.

Nelson-Barber, Sharon, and Elise Trumbull Estrin. *Culturally Responsive Mathematics and Science Education for Native Students*. San Francisco, Calif.: Far West Laboratory for Educational Research and Development, 1995.

Nelson-Barber, Sharon, and Jean Mitchell. "Restructuring for Diversity: Five Regional Portraits." In *Diversity in Teacher Education*, edited by Mary E. Dilworth, pp. 229–62. San Francisco: Jossey Bass Publishers, 1992.

Nunes, Terezinka. "Ethnomathematics and Everyday Cognition." In *Handbook of Research on Mathematics Teaching and Learning*, edited by Douglas A. Grouws, pp. 557–74. New York: Macmillan Publishing, 1992.

Oakes, Jeannie. *Multiplying Inequalities: The Effects of Race, Social Class, and Tracking on Opportunities to Learn Mathematics and Science*. Santa Monica, Calif.: Rand Corporation, 1990.

Palincsar, Annemarie Sullivan. "Less Chartered Waters." *Education Researcher* 18, no. 4 (1989): 5–7.

Philips, Susan Urmston. *The Invisible Culture: Communication in Classroom and Community on the Warm Springs Indian Reservation*. New York: Longman, 1983.

Pimm, David. *Speaking Mathematically: Communication in Mathematics Classrooms*. London: Routledge & Kegan Paul, 1987.

Rauff, James V. "My Brother Does Not Have a Pickup: Ethnomathematics and Mathematics Education." *Mathematics and Computer Education* 30, no. 1 (1996): 42–50.

Scheurich, James Joseph, and Michelle D. Young. "Coloring Epistemologies: Are Our Research Epistemologies Racially Biased?" *Educational Researcher* 26, no. 4 (1997): 4–16.

Schoenfeld, Alan H. "Problem Solving in Context(s)." In *The Teaching and Assessing of Mathematical Problem Solving*, edited by Randall I. Charles and Edward A. Silver, pp. 82–92. Vol. 3, *Research Agenda for Mathematics Education*. Reston, Va.: Lawrence Erlbaum Associates and National Council of Teachers of Mathematics, 1988.

Secada, Walter G. "Race, Ethnicity, Social Class, Language, and Achievement in Mathematics." In *Handbook of Research on Mathematics Teaching and Learning*, edited by Douglas A Grouws, pp. 623–60. New York: Macmillan Publishing Co., 1992.

———. "Understanding in Mathematics and Science" *Principled Practice in Mathematics and Science Education* 1 (spring 1997): 8–9.

Swisher, Karen, and Donna Deyhle. "Styles of Learning and Learning of Styles: Education Conflicts for American Indian/Alaskan Native Youth." *Journal of Multilingual and Multicultural Development* 8, no. 4 (1987): 345–60.

Teacher Panel. "Remarks of Ten American Indian Teachers at a Day-Long Meeting Held at Far West Laboratory, June 29, 1994." San Francisco: Far West Laboratory for Educational Research and Development, 1994.

Trumbull, Elise, Patricia M. Greenfield, Carrie Rothstein-Fisch, and Blanca Quiroz. *Bridging Cultures between Home and School: A Guide for Teachers.* Mahwah, N.J.: Lawrence Erlbaum Associates, 2001.

Zaslavsky, Claudia. *Africa Counts: Number and Pattern in African Culture.* Boston: Prindle, Weber & Schmidt, 1973.

———. "Symmetry in American Folk Art." *Arithmetic Teacher* 38 (September 1990): 6–12.

Teaching Mathematics to American Indian Students

A Cultural Approach

2

David M. Davison

Many educators argue that mathematics is "culture free," that the mathematics curriculum not only is the same for all students but is to be presented to all students in the same way. Bishop (1988), however, believes that mathematics is a cultural product. Culture, as he uses the term, is understood as a set of beliefs and understandings that serves as a basis for communication among a group of people. Both Bishop and D'Ambrosio (1985) have indicated ways in which school mathematics carries particular values that may or may not align with sociocultural contexts. They conclude, in essence, that the alignment of mathematics curricula with culture is necessary if learning is to be enhanced. We observe that the typical textbook format used in American schools is based on a white western European cultural view. When mathematics instruction occurs in only one cultural context, whether dominant or not, students from other cultures are placed at a learning disadvantage.

The problem, then, is the development of culturally rich examples to be used with learners from a variety of backgrounds. Bishop (1988) indicates that the cultural and ethnic backgrounds of students are rich resources from which mathematics concepts can be developed. In fact, the mathematics curriculum can be enriched if teachers working in a cultural environment different from their own are aware of and use mathematically oriented situations in which their students participate outside school. Thus, teachers can select culturally oriented learning activities that can be used in the study of appropriate mathematics topics. Naturally, such learning activities would need to be culturally valid and educationally sound. Cultural validity can be ascertained by seeking information from leaders in the native culture and other native authorities. We should then expect the learning of mathematics to be more effective when embedded in culturally relevant contexts.

APPLICATION TO AMERICAN INDIAN STUDENTS

In a recent survey of twenty-four middle school students in one Crow Indian community, I found that all the students participated in at least one traditional cultural activity. After receiving prompts, they could make connections between the cultural activities and school mathematics. However, the connections were not obvious to them, and they did not regard the cultural activities as intrinsically mathematical. Part of the problem is that the students live in two worlds—the world of their native culture and the world of the dominant culture (as defined by their schooling). Students who attend schools on reservations, seem especially, to have difficulty bridging the two cultures. These students saw mathematics as a bridge to the teenage culture, with little or no relevance to their native culture. When asked how they saw mathematics as affecting their lives, they responded in a manner very similar to that of non-Indian students in the same age group: as a means to getting a good job, to help in buying a car, to assist with building or making things, and the like. Perhaps the most insightful example was provided by the student who indicated that "math helps you figure out how far you can drive on a tank of gas."

These students agreed that using such activities to provide appropriate illustrations for mathematics lessons is helpful, but clearly the concept of relevance for these students means drawing from the stimuli that have the greatest meaning in their lives, the same stimuli that influence most students of this age group—sports, hunting, and music. The implication is that the curriculum should include not only illustrations from the students' native culture but also examples representative of the students' interests.

These Crow Indian students are committed to such Crow cultural activities as hand games, arrow throws, star quilting, beadwork, dancing competitions, teepee raising, and making fry bread. The inclusion of culturally relevant situations in the mathematics curriculum helps make the curriculum more sensitive to the needs of all students, gives students more of a sense of purpose in studying mathematics, and thereby helps raise their mathematics performance. Many activities that are inspired by a knowledge of American Indian cultures could be used as a means of teaching mathematics to American Indian students. The study of geometric patterns in the elementary school can draw from many activities that apply the native culture to the mathematics curriculum. The purpose of this paper is to show how the mathematics curriculum in general and the topic of symmetry in particular can be enriched by the inclusion of such activities.

CULTURALLY RELEVANT APPROACHES FOR AMERICAN INDIAN STUDENTS

Culture-Independent Illustrations

Mary Baratta-Lorton's (1976) *Mathematics Their Way* appears to have proved successful with primary-school-aged children from diverse cultural backgrounds. Its success is linked to the use of materials found in most homes and the forging of connections between the familiar objects and the mathematics concepts. For this reason the program has been well received in American Indian classrooms. Similarly, Gene Maier's (1985) "Math and the Mind's Eye" focuses on the use of culturally sensitive instructional strategies, with particular attention to the forming of mental images through visual connections. More recently, modular programs such as AIMS (Activities Integrating Mathematics and Science) and GEMS (Great Explorations in Math and Science) provide opportunities for diverse learners. Even though these programs are appropriate for use by all students, they appear to have been used very effectively by teachers of American Indian intermediate- and middle-grades students.

Geometric Illustrations

American Indian cultures are replete with examples that relate directly to the mathematics curriculum for all students. The study of symmetry in particular can be enhanced by illustrations from Indian paintings, weaving, beadwork, and star quilting. The investigation of lines of symmetry and rotations in intermediate-level geometry has many applications in Indian crafts.

One way to investigate Indian art designs is to have students create their own, perhaps with the help of such manipulatives as tangrams or pattern blocks. I have done this activity with groups of intermediate-level Indian students, usually directing them only to create a pleasing design with the manipulatives that I have given to them. The result is almost invariably a symmetrical pattern representative of Indian art. It would appear that the informal geometric knowledge demonstrated by these students could be a powerful aid in their development of mathematical concepts. Both tangrams and pattern blocks can prove very effective in developing fraction concepts. For example, using the pattern-block hexagon to represent a unit, students can be asked to find the area of a design that they have created.

Perhaps the most widely known application of Indian craft work is that presented by Marina Krause (1983) in *Multicultural Mathematics Materials.* An example of a Navajo art design is presented in figure 2.1. This design can be used to investigate not only lines of symmetry but also translation and rotation transformations. Her book includes many other designs that are applications of geometry.

Fig. 2.1. Navajo eye-dazzler design
From *Multicultural Mathematics Materials.* © 1983 by the National Council of Teachers of Mathematics

Claudette Bradley's (1975) "Native American Loom Beadwork Can Teach Mathematics" is rich with examples of symmetrical geometric patterns (see fig. 2.2). She also indicates that students must follow problem-solving procedures in executing loom beadwork. Loom beadwork is also helpful in the exploration of other mathematics topics, including fractions. For example, students can

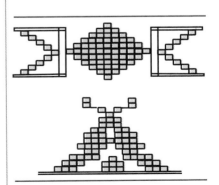

Fig. 2.2. An example of Native American beadwork

determine the fraction of the number of beads of given colors in a design or the ratio of the number of beads in a given pattern to the number of beads in the total design. The activity can be made relevant to American Indian students by telling them who created the beadwork (or better still, having the beadworker make the presentation) and by allowing the students to see the lesson in the context of their whole lives.

Another geometric application is the creation of the star quilt. Even though this task uses high school mathematics, many of the practitioners of the art of star quilting have not studied mathematics at the secondary level. In fact, as in the other examples presented here, the geometric knowledge of the practitioners is rather informal. The star quilt is a rich example to use in the study of symmetry. It has eight lines of symmetry and rotational symmetry of order eight. Moreover, students can create their own visual representation of a quilt using a set of congruent rhombuses (with vertex angles of 45 degrees) as a template (see fig. 2.3).

Fig. 2.3. A star quilt

As with the beadwork example, investigating the star quilt is more meaningful if the students are shown a star quilt, preferably by the person who created it. Doing so helps the students see the informal mathematics of their culture as important.

Curricular Implications

These and other geometry-based illustrations of mathematics used by American Indians can influence the curriculum in one of several ways. First, they can be used to add a cultural dimension to the mathematics curriculum. These cultural examples serve to illustrate topics in the existing curriculum and can fulfill a cross-cultural function in teaching non-Indian students. For Indian students, such examples may provide an opportunity to find some meaning in the mathematics curriculum.

Second, these examples can form part of an integrated presentation of the particular mathematics topic. For example, unit plans could be redesigned to show Indian craft work as an integral part of a unit on symmetry. In essence, this would require adapting the traditional textbook illustrations for the unit to include examples of such art and craft designs.

Third, many educators note that American Indian learners respond well to a thematically integrated curriculum. Accordingly, such topics as the star quilt or the loom beadwork would have applications not only to mathematics but also to a variety of other subjects. An integrated approach is one way of responding to the concern that American Indian students do not find meaning in the curriculum. Again, this approach calls for a substantial investment of time and resources if the integrated curriculum is not simply to be an appendage to the traditional curriculum. In fact, in personal conversations, many teachers of Native American students have remarked that when they introduce a cultural topic in the classroom, they invite an exponent of the cultural art to make the presentation. This step enhances the credibility of the cultural activity and forges a connection between the culture of the community and the culture of the school. It also helps avoid violating any cultural taboos—for example, the creation of objects that have religious significance.

Number-Sense Illustrations

Many examples from the native culture involve the use of games and cultural events. Hand games, Indian dancing, and scoring arrow throws, for example, certainly provide practice in simple arithmetic operations. Including such examples in the mathematics curriculum without investigating the problem-solving implications uses the native culture in a cosmetic way only. I indicated earlier that the learning activities selected need to be culturally appropriate and educationally sound.

Baking Indian fry bread offers a realistic application of ratios when the recipe is expanded for the whole class. Suppose the given fry bread recipe (see fig. 2.4) is for eight people. How would the recipe need to be altered if twenty people were to eat the fry bread? The lesson is more dramatic when the class actually makes and eats the fry bread.

Indian Fry Bread

2 cups flour

1 teaspoon salt

2 teaspoons baking powder

1/2 cup dried milk

1 tablespoon sugar

1 cup warm water

Mix ingredients together. Pat out the dough into a cake eight or nine inches in diameter and one-half inch think, cut it into wedges, and slit each wedge in the center. Fry the wedges in vegetable oil that has been heated to just below the smoking point. Work quickly so that each piece is browned but not burned.

Fig. 2.4. A recipe for Indian fry bread

These examples illustrate that traditional school mathematics can be linked with the informal mathematics of the culture. The challenge for teachers of American Indian students is to look for these connections so that they make sense both in the context of the culture and in the context of the school mathematics curriculum.

CONCLUSION

Students are participants in two cultures—the culture of the home and the culture of the school. American Indian students see little connection between these two cultures; consequently, many potentially rich situations from the native culture are lost to the school. This outcome is particularly true in mathematics, where Native American students feel alienated from the mathematics curriculum. I believe that these students would see more purpose in their school mathematics if it were more closely linked with appropriate aspects of their culture.

I have suggested that the use of cultural situations can improve the learning of mathematics by American Indian students in several ways. The use of familiar situations is one way of helping students attach meaning to the concepts of the school mathematics curriculum. When mathematics ideas from the culture are used as a basis for developing academic mathematics, students value their cultural heritage more. The integration of the students' experiential mathematics and with their school mathematics will help them make connections that they have not previously made. All these strategies discussed in this paper represent ways in which Native American students can overcome their feelings of alienation from school mathematics. The result of implementing these strategies should be improvement in the mathematics performance of American Indian students.

REFERENCES

Baratta-Lorton, Mary. *Mathematics Their Way*. Menlo Park, Calif.: Addison-Wesley Publishing Co., 1976.

Bishop, Alan J. *Mathematical Enculturation*. Dordrecht, Netherlands: Kluwer Academic Publishers, 1988.

Bradley, Claudette. "Native American Loom Beadwork Can Teach Mathematics." Unpublished manuscript, 1975. Available in Harvard University Library.

D'Ambrosio, Ubiritan. *Socio-Cultural Bases for Mathematics Education*. Campinas, Brazil: Unicamp, 1985.

Krause, Marina C. *Multicultural Mathematics Materials*. Reston, Va.: National Council of Teachers of Mathematics, 1983.

Maier, Gene. "Math and the Mind's Eye." Unpublished report of a project funded by the National Science Foundation, 1985.

Standards for Effective Mathematics Education for American Indian Students

3

Ruth Soleste Hilberg

R. William Doherty

Stephanie Stoll Dalton

Daniel Youpa

Roland G. Tharp

Aprimary responsibility of education is to provide a teaching and learning environment in which all students are productive, engaged, and most likely to learn (Tharp, Dalton, and Yamauchi 1994). However, current and traditional teaching strategies have failed to meet the needs of many students. Although alternatives exist, mathematics continues to be taught from the transmission model, in which the teacher transmits knowledge to students by lecturing and modeling problems or problem-solving processes, after which students work individually to solve similar problems (National Council of Teachers of Mathematics [NCTM] 1989, 1991; Wertsch and Toma 1995). As Welch (1978, p. 6), as quoted in NCTM (1991, p. 12), observed,

> In all math classes that I visited, the sequence of activities was the same. First, answers were given for the previous day's assignment. The more difficult problems were worked on by the teacher or the students at the chalkboard. A brief explanation, sometimes none at all, was given of the new material, and the problems assigned for the next day. The remainder of the class was devoted to working on homework while the teacher moved around the room answering questions. The most noticeable thing about math classes was the repetition of this routine.

This model is ineffective for many students, particularly students from certain cultural minority groups that are underrepresented in educational programs and careers that require skill in higher-level mathematics. Consequently, the transmission model for teaching mathematics results in limiting minority groups' access to many of society's opportunities and resources. Although some students flourish under traditional instructional methods, most students do not. The National Research Council's (NRC) report *Everybody Counts* (NRC 1989) asserts that "mathematics is the worst curricular villain in driving students to failure in school. When mathematics acts as a filter, it not only filters students out of careers, but frequently out of school itself" (p. 7).

One group historically excluded from success in mathematics is American Indians. According to Charleston (1994) in the final report of the Indian Nations at Risk task force (INAR), American Indians experience the poorest quality of life of all minority groups in the United States, the highest school dropout rate, and the lowest rate of school attendance. In 1985, standardized-test scores for all schools funded by the Bureau of Indian Affairs showed a dramatic, steady decrease in grade-level performance for American Indian students as they progress through school. Although these students perform near grade level when they enter school, by the time they reach ninth grade they are two grade levels behind majority-culture students, and by twelfth grade they are nearly four grade levels behind (Charleston 1994). The INAR report proposes education as the key to solving the problems confronting American Indians and urges us as a society to "establish education systems that contribute productively to the growth and development of all American societies" (p. 18).

During the past decade mathematics education research has been dominated by two major theoretical trends, constructivism and sociocultural theory (Cobb 1994). The principal distinction between these theories is whether individual processes or social and cultural processes have primary influence on learning and development (Simon 1995; Steffe 1995; Steffe and D'Ambrosio 1995). Constructivist theory—largely derived from the work of Piaget—focuses on the development of the individual and the individual's construction of knowledge. Sociocultural theory—stemming from the work of Vygotsky—emphasizes the influence of social factors on the formation of knowledge. According to sociocultural theory, an individual's mathematical ability is deeply and fundamentally influenced by his or her participation in cultural practices (Cobb 1994). Lewin (1995) contends that the construction of knowledge is seldom, if ever, entirely individual; rather—in both specific content and learning styles such as ways of knowing what is important or worth noticing—it is largely determined by cultural influences.

As part of a national effort to reform mathematics education in the United States, the National Council of Teachers of Mathematics (NCTM) developed professional standards for teaching and learning mathematics (NCTM 1991). NCTM's goals are for *all* students to (1) value mathematics, (2) become confident in their ability to do mathematics, (3) become mathematical problem solvers, and (4) learn to communicate and reason mathematically. Many of NCTM's suggestions for achieving these goals, although based on constructivist theory (NCTM 1991), are also consistent with ideas from sociocultural theory. For example, teachers should guide students in working together to make sense of mathematics, students and teachers should develop common understandings of mathematical ideas, and everyday language should be related to mathematical language and symbols.

The Center for Research on Education, Diversity and Excellence (CREDE) developed five standards for maximizing the school achievement of students at risk of academic failure and two standards specific to effective education for American Indian students. CREDE's Standards for Effective Teaching and Learning (Tharp 1997), although expressed in the language of sociocultural theory, have been extracted from a long line of research in a wide range of settings on the development and evaluation of educational programs for all cultural and social groups. The standards represent a consensus in education research and theory on how to improve teaching and learning for all students and are agreed on by educational researchers and program developers across theoretical domains (Tharp 1988, 1989, 1991, 1994; Tharp et al. 1984). Although disagreements arise on other points and differences are evident in emphasis and degree, remarkable similarities can be found in the recommendations of researchers who have studied programs for all cultural groups.

CREDE's standards, discussed below, are not comprehensive; rather, the assumption is that the acquisition of additional data over time will generate additional or refined consensus standards. For now, these standards are adequate and if enacted will produce a great surge forward in cognitive development, school achievement, and school climate.

CREDE's first standard for effective teaching and learning is to "facilitate learning through joint productive activity among teachers and students"(Tharp 1997, p. 6). Learning is optimized when experts and novices engage in dialogue while working together for a common product or goal (Tharp 1997). Children naturally and often effortlessly develop into competent members of their families and communities by engaging in dialogue and joint activity with more-experienced members. Schools do not typically teach in this manner. This standard asserts that learning is most effective when teachers and students work jointly to solve practical, real-world problems. In such activity, teachers connect academic concepts to everyday concepts, facilitating students' construction of new knowledge from previous knowledge and experiences. Joint productive activity also allows the creation of a common context of experience within the school itself, which is often necessary to learning when the teacher and the students are not of the same background.

Similarly, the NCTM Standards assert that a person creates knowledge in the course of purposeful activity that uses the informal knowledge of the participants. The NCTM Standards also advocate that mathematics classrooms focus on problem solving involving small groups or an entire class working cooperatively and that activities grow out of genuine problem situations in which learning is guided by the search to answer questions. Additionally, the mathematics standards for National Board certification (National Board for Professional Teaching Standards 1998) maintain that accomplished mathematics teachers "provide opportunities for students to talk with each other and work together in solving problems, recognizing that students' justifications and contributions to a group enhance their own learning" (p. 36).

The National Board's mathematics standards provide an interesting example of joint productive activity in which a teacher engages elementary-grades students in the construction of a bar graph based on their preferences for flavors of ice cream. First, each student places a Post-it note on a prepared grid on the chalkboard, and then the students make a human bar graph using classroom floor tiles as a grid and forming lines corresponding to their favorite ice-cream flavors. In this clever activity each student contributes to the creation of a graph that becomes the product of this joint activity. This activity might usefully be modified to make it more culturally relevant to American Indian students, perhaps by basing it on favorite local desserts or traditional meals or on favorite community events or holidays.

Davison (1992) also provides an excellent example of a joint productive activity in which Rosalie Bearcrane, a bilingual teacher at Crow Agency School, taught students to use ratios by dividing a map of the reservation into six rectangles, assigning each portion to a group in the class. Each group enlarged its portion of the map using a scale of 3:1 after which students constructed a plaster relief model of the enlargement. The finished products were combined to form a table-sized relief map of the reservation. In addition to teaching ratios, this joint productive activity motivated students while incorporating both knowledge of the local environment and mathematics.

STANDARD 1: JOINT PRODUCTIVE ACTIVITY— TEACHER AND STUDENTS PRODUCING TOGETHER

STANDARD 2: LANGUAGE DEVELOPMENT— DEVELOPING LANGUAGE ACROSS THE CURRICULUM

CREDE's second standard is to develop competence in the language of instruction across the curriculum. This standard is supported by recent findings in cognitive and education research of strong links among language, cognitive growth, and academic achievement (Tharp 1997). Proficiency in the language of instruction, essential for high academic achievement, must be the first and primary goal of teaching and learning. Language development in both informal and academic domains and in problem solving is the overarching goal for the entire school day. This goal also applies to the specialized language of mathematics. Effective mathematics learning depends on the ability to speak "mathematically," that is, in the language mathematicians use, just as academic achievement across the curriculum is dependent on the mastery of the language of instruction.

Most teachers rely heavily on verbal analytic teaching methods and the use of verbal forms of instruction. Therefore, cultural groups that do not emphasize verbal analytic problem solving are at a disadvantage in school. Additionally, the ways of using language that predominate in schools, such as ways of asking and answering questions or challenging claims, are frequently unfamiliar to language-minority and other at-risk students. The courtesies and conventions of conversation are among the most powerful differentiating elements of culture. For example, a discussion pattern involving long, patient turn taking has been the norm in American Indian meetings, from the earliest recorded powwows to the deliberations of the contemporary American Indian Church. When classroom discussion patterns are incompatible with the patterns American Indian students use in the home and community, many Native American children develop patterns of short answers, interruptions, and silence. By high school, these patterns calcify into a repertoire of hostile and resentful behaviors (Greenbaum and Greenbaum 1983). The result in the classroom is that children's speech is brief, simple, infrequent, or unconventional for the classroom, an outcome that often leads to a school diagnosis of "low verbal ability," even for children who in other settings are highly verbal. However, by learning about culturally based ways of communicating and by incorporating them into classroom discourse and instruction, teachers can encourage children's language and communication strengths.

Considerable support for CREDE's Language Development standard can be found in NCTM's *Curriculum and Evaluation Standards for School Mathematics* (1989)(*Curriculum and Evaluation Standards*), which maintains that communication helps students

> construct links between their informal, intuitive notions and the abstract language and symbolism of mathematics.... Young children are active, social individuals. Much of the sense they make of the world is derived from their communications with other people. Communicating helps children to clarify their thinking and sharpen their understandings.... [E]ncouraging them [students] to represent, talk and listen, write, and read facilitates meaningful learning. (pp. 26–28)

Additionally, the National Board for Professional Teaching Standards (1998) asserts that accomplished mathematics teachers provide frequent opportunities for students to explain their thinking, make generalizations, justify their conclusions, and relate their attitudes and feelings about mathematics.

Moore (1994) provides an illustration of how language may affect the mathematics learning of American Indians. In the Navajo language the first noun that appears in a sentence is assumed to be the subject of the sentence. Moore attempted to test how this might affect the mathematics problem-solving abilities of Navajo university students. He wrote the sentence "The water was drunk by the girl" on the chalkboard, after which the students laughed and explained that the sentence implied that the water drank the girl. The structure of these students' native language influenced their interpretations of the sentence. Similar misinterpretations or mistranslations would most certainly affect the mathematical problem solving of Navajo students, particularly when solving complex, abstract, or contrafactual mathematical problems.

Moore (1982) also reports a study in which a graduate student demonstrated the difficulty Navajo students have when translating word problems based on hypothetical situations using the word *if*. The Navajo language does not have a word that directly translates to *if*. The researcher found that Navajo students paraphrased if-then statements differently from the way native English-speaking students do, often substituting the Navajo words for *when, and,* or *or.* This and other translation difficulties could present Navajo students who are not proficient in English with a distinct disadvantage in mathematics.

CREDE's third standard is to connect teaching and curriculum to students' experiences and to the skills of their home and community. Instruction is more effective when contextualized in students' personal experiences and knowledge (Tharp 1997). Connecting educational content to students' personal lives and the values and traditions of the local community and providing instruction in familiar, everyday contexts about which students have prior knowledge enable students to make sense of instruction and to construct new knowledge accordingly. Three levels of contextualization are possible in schools. At the pedagogical level, patterns are established in the classroom of participation and discourse that are familiar to students at home and in the community. At the curriculum level, materials and skills from home and community serve as the foundation for academic content and the development of school skills. At the policy level, the school itself is contextualized in the community and reflects its beliefs, values, and goals.

The NCTM Standards concur with CREDE's Contextualization standard. An overarching goal of all the NCTM Standards is to integrate mathematics into contexts that give practical meaning to its symbols and processes. NCTM encourages teachers to find ways to assist students for whom the notion of argument is uncomfortable or incompatible with community norms in engaging in mathematical discourse. Teachers are also encouraged to pose tasks that display sensitivity to, and draw on, students' diverse experiences and dispositions. *Professional Standards for Teaching Mathematics* (NCTM 1991) (*Professional Teaching Standards*) asserts,

> Teachers also need to understand the importance of context as it relates to students' interests and experience. Instruction should incorporate real-world contexts and children's experiences and, when possible, should use children's language, viewpoints, and culture. Children need to learn how mathematics applies to everyday life and how mathematics relates to other curriculum areas as well. (p. 146)

Britton (1983) provides an excellent example of a mathematics lesson contextualized for Plains Indian students. An interdisciplinary unit, "Medicine Wheels: The Art and Culture of the Plains Indian," was designed to incorporate art, social studies, and mathematics. Students develop compass, protractor, and ruler skills while making geometric constructions. They construct bisectors,

STANDARD 3: CONTEXTUAL-IZATION—MAKING MEANING BY CONNECTING SCHOOL TO STUDENTS' LIVES

29

parallel lines, and congruent segments and angles and learn about ratio and proportion in the creation of their personal medicine wheels. The mathematics in this well-constructed unit is contextualized in the history and art of Plains Indians.

Schaufele and Srivastava (1995) developed a college algebra course—Earth Algebra—for students at Navajo Community College (NCC) with the goal to "bring college algebra to life, to put it in a context that would provide real situations and familiar circumstances where mathematics could be used to solve meaningful problems and reach informed decisions" (p. 13). The successful Earth Algebra materials were originally developed by Schaufele and Zumoff for students in Georgia, and when Schaufele attempted to teach the course at NCC, the Navajo students did not initially respond with enthusiasm. The focus of the course was on environmental problems, specifically energy consumption and tropical deforestation. Not until the curriculum was modified to emphasize environmental problems involving water conservation, water quality, and air pollution—major concerns of the NCC community—did students begin to react positively. This course was further contextualized by modeling mathematical problem-solving processes after traditional community problem-solving processes associated with the four directions: the problem situation is visualized, a problem-solving plan is developed, a solution is obtained, and finally the solution is checked against the original problem and visualization. The course was contextualized first by modifying the curriculum and building it around problems of interest and concern to the students and then by incorporating traditional problem-solving processes.

STANDARD 4: COGNITIVE CHALLENGE— TEACHING COMPLEX THINKING

CREDE's fourth standard is to "challenge students toward cognitive complexity." A clear consensus exists among researchers that at-risk students require instruction that is cognitively challenging (Tharp 1997). Instruction must include thinking and analysis, not merely rote or repetitive drills to teach basic skills. This precept does not mean omitting the memorization of multiplication tables, but it does mean that instruction for all students must go beyond the instruction and practice of basic skills. At-risk students, particularly those with limited standard English proficiency, as is often true of American Indian students, are often not held to the same standards as mainstream students or students from the majority culture. The logical result is that achievement is hindered. Although often the result of benign motives, the effect is to deny many diverse students the basic requirements for achievement: high academic standards and meaningful assessment that allows for necessary feedback and responsive assistance.

Again, considerable agreement exists between CREDE's and NCTM's standards. The *Professional Teaching Standards* (NCTM 1991, p. 27) states that "defensible reasoning about students must be based on the assumption that all students can learn and do mathematics, that each one is worthy of being challenged intellectually," and that "the teacher of mathematics should create a learning environment that fosters the development of each student's mathematics power by consistently expecting and encouraging students to take intellectual risks" (p. 57).

An excellent example of a cognitively challenging curriculum for American Indians was created for Navajo and Hopi students at a high school in Arizona. The teachers worked with university faculty to develop a program, Bio-Prep, to prepare students for college and careers in science and health (Rist 1992). The Bio-Prep program consists of an accelerated high school curriculum with an emphasis on mathematics and science. For example, in addition to algebra and biology, the ninth-grade curriculum includes a course called Math-Physics that teaches mathematics through problem solving in physics. Students in the Bio-Prep program receive academic support, encouragement, and intense guidance. The program accepts students performing at grade level and requests merely that both students and parents commit to high academic goals and hard work. The cognitive challenge provided by the Bio-Prep program has dramatically increased students' achievement. As a result, the once predominantly Anglo honor society has become overwhelmingly American Indian, students have won an impressive list of awards and scholarships, and many Bio-Prep students have gone on to such schools as the Massachusetts Institute of Technology. Rather than being "filtered out," these students are being filtered into some of the finest universities in the nation.

CREDE's fifth standard is to "engage students through dialogue, especially the Instructional Conversation" (Tharp 1997). Basic thinking skills are most effectively developed through dialogue, through the process of questioning and sharing ideas and knowledge that happens in the "instructional conversation" (Tharp 1997). The instructional conversation is the means by which teachers relate formal, academic knowledge to the students' personal, family, and community knowledge. Teachers who use conversation to provide instruction, like parents in natural teaching, assume that students have something to say beyond the anticipated answers. The teacher listens carefully, makes guesses about intended meanings, and adjusts responses to assist the student. Such conversation reveals the knowledge, skills, and values of the learner, enabling the teacher to contextualize teaching to fit the learner's experience and provide the assistance necessary to move the student forward, thus individualizing instruction.

NCTM's *Curriculum and Evaluation Standards* (1989) is in complete agreement with CREDE that dialogue is the means by which instruction is contextualized:

> The study of mathematics should include numerous opportunities for communication so that students can relate physical materials, pictures, and diagrams to mathematical ideas; reflect on and clarify their thinking about mathematical ideas and situations; relate their everyday language to mathematical language and symbols.... Attending to students' communications about their thinking also gives teachers a rich information base from which they can make sound instructional decisions. (pp. 26, 28)

An example of how instruction can occur within conversation is found in an interview by Franklin (1991) on the Zuni reservation in New Mexico. Teacher Odell Jaramillo describes how she engages students in a highly contextualized discussion and skillfully incorporates mathematical terminology and concepts into experiences familiar to Zuni students:

> Zuni is a place where they have dances all year round and then for another art project I tell them to go watch a dance and then we talk about it—the designs that they saw, that they wore. You know, how did it look—the triangles, the rectangles. What kinds of shapes did they see? Instead of just talking about the dancers and their singing, I talked about their religion. Little boys, especially at kindergarten, they're not really exposed, they're not initiated, so there's a limited amount of what they can say or of what they can do. I also wanted them to get into shapes and knowing the sashes, making them with yarns. We make sashes, belts and stuff like that. And there's the symbol and meaning of the rain cloud and then the sun and

STANDARD 5: INSTRUCTIONAL CONVERSATION— TEACHING THROUGH CONVERSATION

different things. We get back into language. Before there was (written) language there were symbols that people used to talk with and so we show them the patterns that they used for the sun and the rainbows, too. (p. 4)

Moore (1994) used an informal whole-class discussion to contextualize algebra instruction for Navajo college students. American Indian students learn more effectively if taught in a holistic manner (Tharp 1994) in which students are initially presented with a view of the larger context so that the meaning of concepts can be derived from their relationship to that context. Typical mathematics instruction assumes that students will develop an understanding of concepts from their relationships to one another, not the larger context. American Indian students may prefer or require an understanding of the whole picture before proceeding to the details. When Moore taught a method for solving quadratic equations, he initiated the lesson with a discussion of a problem situation in which obtaining a solution required solving a quadratic equation:

For example, I posed the following problem. "If you throw a baseball straight up, when will it be fifty feet high?" There were some sneers and snickers and responses such as, "Well, how fast was the ball thrown?" I endured these remarks and persisted with my initial question until one student observed: "But there will be two times when the ball is fifty feet high, one with the ball going up and a second when the ball comes down." We then discussed models of equations with more than one solution. (p. 10)

CREDE also examined the research literature for additional recommendations that may be distinctive of American Indian education. Two standards were developed that are either explicit or embedded in effective programs for American Indian students. These recommendations are not surprising, since they are closely tied to basic views of children and basic characteristics of child socialization that appear to be shared by American Indian cultures.

STANDARD 6: CHOICE AND INITIATIVE— ENCOURAGING STUDENTS' DECISION MAKING

CREDE's sixth standard is "include activities that are generated and directed by individual students or small groups" (Tharp, Dalton, and Yamauchi 1994, p. 37). The way classrooms are organized affects the level of participation and motivation of American Indian students (Tharp 1997). Because of the high level of autonomy and decision making granted to youth in Native American cultures, American Indian students are more comfortable and more motivated to participate in activities that they generate, organize, or direct themselves. Students even participate more in instructional conversations with teachers in classrooms organized into small student-directed groups. For example, effective American Indian classroom organization might allow for spontaneous instructional conversations among the teacher and up to seven students, as well as scheduled time for more-structured conversations. The teacher may "float" among individuals and small groups or be stationary but approachable (Tharp, Dalton, and Yamauchi 1994). Such a teacher offers responsive instructional conversation as needed while allowing students opportunities to initiate and terminate those conversations. This approach provides an opportunity for teachers to work on language development with each student, to contextualize instruction, and to provide responsive assistance while other students are involved in their own pursuits.

Although NCTM's *Professional Teaching Standards* does not directly address the issue of student choice in the mathematics classroom, it does discuss the importance of a classroom environment in which students have time to ponder and think and in which teachers display "sensitivity to the diversity of students' backgrounds and experiences" (NCTM 1991, p. 27). In the example of American

Indians, sensitivity to students' backgrounds would necessitate respect for the autonomy that parents grant their children and to which the children are accustomed.

In an elementary school program designed to improve mathematics education for American Indian students in northern Minnesota, Wallis (1983) incorporated student choice—which she terms *flexibility*—to enhance learning. In a given unit students choose the number of concepts they wish to focus on and set their own goals for achievement. To master each concept, students visit learning centers at which they engage in work activities and games to develop skill and understanding in the specified concept area. Wallis (1983) reports that allowing student choice in the classroom improved students' attitudes and performance.

STANDARD 7: MODELING AND DEMONSTRATION —LEARNING THROUGH OBSERVATION

CREDE's seventh standard proposes that teachers "include some performance and demonstration" (Tharp, Dalton, and Yamauchi, 1994, p. 37). This standard is based on the tradition of American Indian learning through observation and on the custom of allowing students to develop competence before requiring them to perform publicly (Tharp 1997). This observational learning style is closely tied to the visual learning patterns of American Indian children and their holistic cognitive style (Tharp 1994). Even effective instructional conversations can include demonstration. Rather than detract from conversational opportunity, the inclusion of a modeling or demonstration activity is likely to facilitate conversation. The inclusion of models or demonstrations in lessons increases students' understanding of explanations, especially for students whose proficiency in the language of instruction is limited.

NCTM's *Professional Teaching Standards* (1991) asserts that "modeling mathematical ideas through the use of representations (concrete, visual, graphical, symbolic) is central to the teaching of mathematics. Teachers need a rich, deep knowledge of the variety of ways mathematical concepts and procedures may be modeled" (p. 151).

Most mathematics education typically involves teachers' modeling problem-solving strategies; an example of another type of modeling is found in Koelsch (1994, as cited in Nelson-Barber and Estrin [1995, p. 181]). Fifth-grade teachers at a Navajo school developed a mathematics lesson in which students are presented with an example—a model—of a rug pattern and are then asked to design their own pattern and make calculations on the basis of their designs (this is also an excellent example of contextualization). Students might also benefit from watching the teacher create a design before proceeding to make their own.

CONCLUSION

The Center for Research on Education, Diversity and Excellence has proposed a set of standards that represents a consensus of recommendations for effective teaching and learning. Although these standards are distilled from varied work in minority education, generally in underachieving, culturally and linguistically diverse classes, their alignment with NCTM's Curriculum Standards and Professional Teaching Standards suggests that CREDE's proposals would be equally effective for all students. CREDE's Standards for Effective Teaching and Learning are not discrete principles; rather, they form one holistic view. The instructional conversation is most effective when students and teacher engage in joint productive activity that includes modeling and demonstration and student choice. The instructional conversation is also the means by which instruction is contextualized, the language of instruction is developed, and teachers obtain the feedback necessary to challenge all students cognitively.

Meeting the needs of all students will require extensive reform but how, as Wilson et al. (1996, p. 469) asked, can "a system as stitched into the basic fabric of our society as our schools learn to remake itself around a more ambitious and socially responsible set of goals?" The teaching and learning environment proposed by CREDE's standards is dramatically different from that experienced by today's teachers when they were students. Therefore, essential to the success of reform is an understanding of how teachers can be helped to implement these changes. CREDE, currently engaged in a systemic reform effort in the Zuni Public School District, uses the standards presented here to teach these same standards to teachers. CREDE researchers use instructional conversations, modeling and demonstration, and joint productive activity with parents, other community members, teachers, and school administrators to design contextualized curriculum units and develop a schoolwide system that incorporates all the standards and meets the needs of *all* students.

REFERENCES

Ball, Deborah L. "Teacher Learning and the Mathematics Reforms." *Phi Delta Kappan* 77, no. 7 (1996): 500–508.

Britton, Jeanette. *Medicine Wheels: The Art and Culture of the Plains Indian.* An interdisciplinary unit for seventh-grade students involving art, social studies, [and] mathematics. Seattle, Wash.: Seattle Public Schools, 1983.

Charleston, G. Mike. "Toward True Native Education: A Treaty of 1992." Final report of the Indian Nations at Risk task force. *Journal of American Indian Education* 33, no.2 (1994): 1–56.

Cobb, Paul. "Where Is the Mind? Constructivist and Sociocultural Perspectives on Mathematical Development." *Educational Researcher* 23, no. 7 (1994): 13–20.

Dalton, Stephanie Stoll. "Pedagogy Matters: Standards for Effective Teaching Practice." Santa Cruz, Calif.: Center for Research on Education, Diversity and Excellence, University of California, 1998.

Davison, David M. " Mathematics." In *Teaching American Indian Students*, edited by Jon Reyhner, pp. 214–50. Oklahoma City: Oklahoma Press, 1992.

Franklin, Elizabeth. "Integrating Home and Community into the Curriculum: The Zuni Experience." *Insights into Open Education (*Center for Teaching and Learning, University of North Dakota*)* 23, no. 8 (1991): 2–8.

Greenbaum, Paul, and Susan C. Greenbaum. "Cultural Differences, Nonverbal Regulation, and Classroom Interaction." *Peabody Journal of Education* 61, no. 1 (1983): 16–33.

Koelsch, Nanette. *Chinle Public Schools Portfolio Assessment Implementation Manual.* San Francisco: Far West Laboratory, 1994. Quoted in Sharon Nelson–Barber and Elise Estrin, *Bringing Native American Perspectives to Mathematics and Science Teaching* (San Francisco: Far West Laboratory for Education Research and Development, 1995), p. 181.

Lewin, Philip. "The Social Already Inhabits the Epistemic: A Discussion of Driver; Weed, Cobb, and Yackel; and von Glasersfeld." In *Constructivism in Education*, edited by Leslie P. Steffe and Jerry Gale, pp. 423–32. Hillsdale, N.J.: Lawrence Erlbaum Associates, 1995.

Moore, Charles G. *The Navajo Culture and the Learning of Mathematics.* Washington, D.C.: National Institute of Education, 1982.

———. "Research in Native American Mathematics Education." *For the Learning of Mathematics* 12, no. 2 (1994): 9–14.

National Board for Professional Teaching Standards. *Middle Childhood through Early Adolescence/Mathematics Standards.* Southfield, Mich.: National Board for Professional Teaching Standards, 1998.

National Council of Teachers of Mathematics (NCTM). *Curriculum and Evaluation Standards for School Mathematics*. Reston, Va.: NCTM, 1989.

———. *Professional Standards for Teaching Mathematics*. Reston, Va.: NCTM, 1991.

National Research Council, Mathematical Sciences Education Board. *Everybody Counts: A Report to the Nation on the Future of Mathematics Education*. Washington, D.C.: National Academy Press, 1989.

Rist, Marilee C. "Miracle in Arizona." *Executive Educator* 14, no. 7 (1992): 32–35.

Schaufele, Christopher, and Rivindra Srivastava. "Earth Algebra: Real-Life Mathematics in Navajoland." *Journal of Navajo Education* 12, no. 2 (1995): 12–15.

Simon, Martin A. "Reconstructing Mathematics Pedagogy from a Constructivist Perspective." *Journal for Research in Mathematics Education* 26, no. 2 (1995): 114–15.

Steffe, Leslie P. "Alternative Epistemologies: An Educator's Perspective." In *Constructivism in Education*, edited by Leslie P. Steffe and Jerry Gale, pp. 489–523. Hillsdale, N.J.: Lawrence Erlbaum Associates, 1995.

Steffe, Leslie P., and Beatriz S. D'Ambrosio. "Toward a Working Model of Constructivist Teaching: A Reaction to Simon." *Journal for Research in Mathematics Education* 26, no. 2 (1995): 146–59.

Tharp, Roland G. "Cultural Diversity and Treatment of Children." *Journal of Consulting and Clinical Psychology* 59, no. 6 (1991): 799–812.

———. "From At-Risk to Excellence: Research, Theory, and Principles for Practice." Santa Cruz, Calif.: Center for Research on Education, Diversity and Excellence, University of California—Santa Cruz, 1997.

———. "Intergroup Differences among Native Americans in Socialization and Child Cognition: An Ethnographic Analysis." In *Cross-Cultural Roots of Minority Child Development*, edited by Patricia Greenfield and Rodney Cocking, pp. 87–105. Hillsdale, N.J.: Lawrence Erlbaum Associates, 1994.

———. "Psychocultural Variables and Constants: Effects on Teaching and Learning in Schools." *American Psychologist* 44, no. 2 (1989): 349–59.

———. *Rousing Minds to Life: Teaching, Learning, and Schooling in Social Context*. New York: Cambridge University Press, 1988.

Tharp, Roland G., Stephanie Dalton, and Lois A. Yamauchi. "Principles for Culturally Compatible Native American Education." *Journal of Navajo Education* 11, no. 3 (1994): 33–39.

Tharp, Roland G., Cathie Jordan, Gisela E. Speidel, Kathryn H. Au, Thomas W. Klein, Roderick P. Calkins, Kim C. Sloat, and Roland Gallimore. "Product and Process in Applied Developmental Research: Education and the Children of a Minority." In *Advances in Developmental Psychology*, edited by Michael E. Lamb, Ann L. Brown, and Barbara Rogoff, pp. 91–141. Hillsdale, N.J.: Lawrence Erlbaum Associates, 1984.

Wallis, Patricia. "Holistic Learning—a Must with American Indian Students." *Momentum* 14, no. 1 (1983): 40–42.

Welch, Wayne. "Science Education in Urbanville: A Case Study." In *Case Studies in Science Education*, edited by R. Stake and J. Easley. Urbana, Ill.: University of Illinois, 1978, p 6. Quoted in National Council of Teachers of Mathematics (NCTM), *Professional Standards for Teaching Mathematics* (Reston,Va.: NCTM, 1991, p. 1).

Wertsch, James, and Chikako Toma. "Discourse and Learning in the Classroom: A Sociocultural Approach." In *Constructivism in Education*, edited by Leslie Steffe and Jerry Gale, pp. 159–74. Hillsdale, N.J.: Lawrence Erlbaum Associates, 1995.

Wilson, Suzanne M., Penelope L. Peterson, Deborah L. Ball, and David K. Cohen. "Learning by All." *Phi Delta Kappan* 77, no. 7 (1996):468–76.

Investigating the Correspondence between Native American Pedagogy and Constructivist-Based Instruction

4

Surprisingly little attention has been given to the teaching methods used in teaching ethnic minority students in this country, particularly when the notion of culturally relevant curriculum materials has been around as long as it has. It is as if we have been able to recognize that there are cultural differences in what people learn, but not in how they learn.
—Susan Urmston Philips, *The Invisible Culture*

The preceding quotation serves as a lens through which this review of literature is focused. In it, Philips (1983) succinctly identifies a serious issue—the importance of culturally compatible teaching methods. This issue grounds our investigation of the pedagogical compatibility between Native American ways of teaching and constructivist-based instruction. In this article, both Native American and constructivist pedagogy will be described and the commonalities of the two will be discussed.

Although we have proposed that Native Americans have a shared way of teaching, it is important to recognize that they represent more than 280 different tribal groups (Butterfield 1983; MacDonald 1989) and that each tribal group possesses varying linguistic, cultural, social, political, and economic dimensions. A consideration of such diversity forces us to be cautious not to overgeneralize about consistencies among tribes (Brassard and Szaraniec 1983). However, many studies point out that notable consistencies do exist among Native peoples (Tharp and Yamauchi 1994; Cahape and Howley 1992; Indian Nations at Risk Task Force 1992). This article, like those cited above, identifies commonalities, specifically informal and formal teaching practices, shared by Native American teachers across tribes. These practices constitute what is referred to in this article as *Native American pedagogy*.

In order to gain an understanding of traditional Indian teaching, it is important to discuss the Native American perspective on the learner's role. Tafoya (1982) helps illuminate this perspective by relating an incident of a Navajo elder responding to a young boy's query about why it snows in Montezuma Canyon. The elder responded by telling him a story about a boy who discovered a strange flaming object:

Judith Elaine Hankes
Gerald R. Fast

PERSPECTIVES ON THE LEARNER'S ROLE

37

They [the Holy People] would not allow him to keep even a part of it, but instead put him to a series of tests. When he was successful at these tests, they promised they would throw all of the ashes from their fireplace into Montezuma Canyon each year. "Sometimes they fail to keep their word, and sometimes, they throw down too much; but in all, they turn their attention toward us regularly, here in Montezuma Canyon." (P. 27)

When the boy heard the story, he accepted the explanation of why it snowed in Montezuma Canyon but then wanted to know why it snowed in Blanding, another Navajo area. The old man quickly replied, "I don't know. You'll have to make up your own story for that." (P. 28)

Tafoya's explanation of this interchange situates the learner as an active participant, not merely a passive recipient of knowledge (p. 28):

This is very much a part of Native American teaching: that one's knowledge must be obtained by the individual ... [and a] gaining of that knowledge does not come from only listening to elders, or seeing what others have done.... The seeker must open up himself.... The insights and comprehensions must be achieved internally.

Others (Leavitt 1983; Scollon and Scollon 1981) have investigated both the social and cognitive aspects of storytelling as well as other traditional Native American teaching practices and, like Tafoya, have concluded that abstracting basic rules and principles is left to participants according to their perspectives and levels of experience.

This conclusion situates the learner not as a dependent student but as an autonomous learner. Understanding this perspective becomes vitally important when we are analyzing the principles of Native American pedagogy. The learner's autonomy is covertly embedded in indirect, cooperative, sense-making, culturally situated, and time-generous instruction.

Constructivist perceptions of the learner and the learner's behaviors parallel those described above and conflict with the perspectives of the dominant culture. Historically, in American classrooms the learner has been viewed as "an empty vessel" or "blank slate," and learning has been thought to be a repetitious "mimetic" activity (Jackson 1986). In contrast, constructivism emphasizes the learner as an active maker, or constructor, of meaning (Glatthorn 1994). In _A Case for Constructivist Classrooms_, a source frequently referenced in this article, Brooks and Brooks (1993, p. 16) explain that

the constructivist vista is far more panoramic and elusive. Deep understanding, not imitative behavior, is the goal.... In the constructivist approach, we do not look for what students can repeat, but for what they can generate, demonstrate, and exhibit.

COMPARING PEDAGOGIES

In the following section, we review four research studies detailing the instruction of Indian children by Indian teachers. Although the studies share attributes common to traditional Native American instruction, such as the autonomous role of the learner, each study was selected to exemplify a particular component or principle of such pedagogy, and each study is identified by the principle it exemplifies. Constructivist correspondence with each principle is also discussed. The pedagogical principles considered are (1) the teacher as facilitator: guiding rather than telling; (2) sense-making instruction; (3) problem-based instruction with problems situated in the culture and lived experiences of the learner; (4) cooperative rather than competitive instruction; and (5) time-generous rather than time-driven instruction.

Principle 1: The Teacher as Facilitator—Guiding Rather Than Telling

Eriks-Brophy and Crago (1993) analyzed classroom discourse in six Inuit-taught kindergarten and first-grade classes in northern Quebec. The findings of this study suggest that unlike mainstream instruction based on behaviorist learning theory, which typically organizes discourse around elicitation sequences initiated and controlled by the teacher (Cazden 1988; Silliman and Wilkinson 1991), Inuit instruction facilitated group responses and peer modeling to decentralize intervention and de-emphasize the authoritarian role of the teacher. The emphasis in the Inuit classrooms was on listening to others as opposed to responding and performing individually, and this focus allowed the teachers to take advantage of peer models to explore emerging concepts as well as to rethink misconceptions. In their interviews, the teachers described their role and goals in facilitating peer exchanges—

- "to encourage my students to get along and help each other";
- "that my students learn to cooperate";
- "that my students respect each other";
- "to keep all the children equal";
- "to be a good example to my students."

The researchers suggested that this pattern of interaction is culturally congruent with the larger Inuit society, in which topics of conversation are not controlled by individual speakers and verbal interactions are typically focused on the general audience and do not tend to spotlight individual participants. They concluded that teachers avoided singling students out for evaluation, praise, or correction in front of their peers and were careful not to emphasize or spotlight individual performance in the public arena, allowing students to participate equally in classroom exchanges without pressure or loss of face. In this way, students were able to take greater responsibility for their own learning and the progress of the group. At the same time, they learned central Inuit values concerning the importance of group cooperation, the equality of all group members, and respect for others.

The findings of this study are consistent with those of other Native American education studies (Erickson and Mohatt 1982; Foster 1989; Philips 1983; Lipka 1991; Scollon and Scollon 1981). The role of the teacher as facilitator promotes both autonomous and cooperative learning: students take greater responsibility for their own learning.

Collins, Brown, and Newman (1989) note that in constructivist classrooms the teacher coaches and "scaffolds," observes the learner, and offers hints and feedback to guide her or his thinking. This guidance includes encouraging the student to reflect on, and talk about, that thinking as well as to compare it with the thinking of others. Constructivist classrooms are not dominated by textbook-oriented "teacher talk"; rather, they are environments of collaborative problem solving. Brooks and Brooks (1993, p. 103) caution against domination by the teacher and the textbook:

Conscientious students who are acculturated to receiving information passively and awaiting directions before acting will study and memorize what their teachers tell them is important. Robbing students of the opportunity to discern for themselves importance from trivia can evoke the conditions of a well-managed classroom at the expense of a transformation-seeking classroom.

Principle 2: Sense-Making Instruction; and *Principle 3:* Problem-Based Instruction with Problems Situated in the Culture and Experiences of the Learner

In a case study of a Yup'ik teacher, Lipka (1991) proposes that Eskimo teachers should teach Eskimo students. He states that ethnicity is not merely a classroom variable but determines the actual style of interaction and the relationship between students and teachers. He argues that for teachers to teach effectively, they not only must possess an in-depth understanding of the content in the culture but also must share culturally compatible communication styles and values. Lipka grounds his proposal on cognitive learning theory (p. 205):

> Research on minority and indigenous school-aged students reveals a "relational" cognitive style. [This] style recognizes the importance of the whole and the context, as opposed to an "analytical" cognitive style, which is abstract and decontextualized (Cohen 1969).... The non-Yup'ik teacher states in a linear manner, first this, and then this (Good & Brophy 1987), prior to doing anything. The verbal messages are decontextualized from the content.... Differences in ordering of introductory statements between Anglo and Yup'ik teachers are not mere happenstance; they are culturally grounded in Yup'ik and mainstream American culture. Activities [in indigenous classrooms] begin without the customary lengthy verbal introduction Anglos expect. This suggests differences in cognitive ordering and structuring.

Lipka illuminates the intentionality of Native American instruction by explaining specific practices. For example, Native American teachers will ignore requests for procedural or locational assistance (e.g., directing students where to sit, stand, move, and so on) so as to avoid reinforcing dependence on verbal instruction during lessons that call for observation. They also allow students to move from their seats to get a closer look at what other students might be doing. In these ways, the teacher shares instruction with the students and builds group solidarity.

Another Native American teaching practice is to avoid correcting students. In Yup'ik pedagogy, feedback takes the form of affirming what students know instead of telling them directly that they are wrong. Underlying this practice is the belief that each student is capable of learning when allowed to perform in his or her comfort zone.

In the Yup'ik classroom of Lipka's case study, instruction was relaxed, almost informal, and reminiscent of lessons taught by elders. The teacher involved the students in content exploration by situating it pedagogically within their lived experiences. Among the examples of situated pedagogy described by Lipka was an incident in which the teacher invited the students to sit on the floor with him. Some chose to, and some chose not to do so. The teacher neither coaxed nor persuaded the students. From a Yup'ik cultural perspective, the teacher was respecting the individual and reinforcing group harmony.

Likewise, the content of the lesson—an art lesson simulating the stretching of a beaver skin—was situated in the Yup'ik culture and required a special understanding by the teacher, who had to perceive the lesson not only as an art activity but also as a lesson about survival, patience, care, and doing things properly.

To summarize, for instruction to be culturally sensitive, content and pedagogy must be culturally congruent. Other researchers reporting similar findings include Gilliland (1992), Macias (1989), Ross (1989), and Swisher and Deyhle (1989).

Constructivism also emphasizes situated and contextualized learning. In constructivist classrooms, students carry out tasks and solve problems that resemble those in the real world (Glatthorn 1994). Instead of doing exercises out of context, the student becomes engaged with contextualized problems that allow the learner to connect prior knowledge to new knowledge and transfer new knowledge and understanding to real situations.

Principle 4: Cooperative Rather Than Competitive Instruction: Cooperative Self-Determination

In *The Invisible Culture,* Philips (1983) provides a detailed description of how classroom communication patterns of Indian children parallel adult communication patterns on the Warm Springs Indian Reservation in Oregon. The thought-provoking findings of this study include the following:

> Indian students generally make less effort than Anglo students to get the floor in classroom interaction. They compete with one another less for the teacher's attention, and make less use of the classroom interactional framework to demonstrate academic achievement. (P. 108)

> It is generally the case that turns at talk are more evenly distributed in Indian classrooms. (P. 113)

> In the Indian classroom it is common for children who had demonstrated the ability to answer correctly a particular question in one instance to refrain from even trying to answer the same question the next time it was raised. (P. 113)

> The children are reared in an environment that discourages drawing attention to oneself by acting as though one is better than another. (P. 118)

> Indian student verbal participation in group projects was not only much greater than in whole class or small-group (teacher directed) encounters, but also qualitatively different. As a rule, one could not determine who had been appointed as leaders of the Indian groups on the basis of the organization of interaction, and when the students were asked to pick a leader, they usually ignored the instructions and got on to the task at hand. (P. 120)

> Indian children demonstrate a strong preference for team games and races … but they show a reluctance to function as leaders in games that require one person to control the activities of others. (P. 122)

> In group projects and playground activities, Indian students were able to sustain infrastructure interactions involving more students for a longer time (compared to white students in the same study) without the interaction breaking down because of conflict or too many people trying to control the talk. (P. 124)

Philips proposes that the noncompetitive, cooperative behavior of Indian children mirrors adult communication patterns. She hypothesizes that the Indian organization of interaction can be characterized as maximizing the control that an individual has over his or her own talk and minimizing the control that a given individual has over others. Furthermore, she explains that Indians are accustomed not to having to appeal to a single individual for permission to speak but rather to determining for themselves whether they will speak. Behavior that might be judged by Anglos to be reticent and insecure would be considered by Indians to be self-determining.

Many studies of North American Indians have noted that overt authority that would interfere with the autonomy of the individual is rarely or never exercised (Basso 1970; Hallowell 1955; Erickson and Mohatt 1982; Spindler and Spindler 1971). Indians are not by nature self-effacing; rather, Philips suggests, this behavior is a consequence of the caretaking patterns of extended families. A child cared for by siblings and cousins is less likely to compete for the attention

41

of a dominant adult and is more likely to attend to the rhythms of group inter-action while maintaining a healthy degree of independence. In school, Indian children probably encounter for the first time the idea that a single individual can be set apart structurally from everyone else in a group, in a role other than that of an observer, and still be a part of the group organization (Philips 1983).

Like the patterns of social interaction described by Philips, the foundational principles of constructivism emphasize the social and cultural mediation of learning, as concisely expressed by Brooks and Brooks (1993, p. vii):

> Constructivism is not a theory about teaching. It's a theory about knowledge and learning. Drawing on a synthesis of current work in cognitive psychology, philos-ophy, and anthropology, the theory defines knowledge as temporary, developmen-tal, socially and culturally mediated, and thus, non-objective. Learning from this perspective is understood to be a self-regulated process of resolving inner cogni-tive conflicts that often become apparent through concrete experiences, collabo-rative discourse, and reflection.

The cooperative learning behaviors described earlier by Philips are rich examples of social mediation and must be regarded with awe and respect. Such behaviors exemplify not only human ways of learning but humane ways of surviving.

Principle 5: Time-Generous Rather Than Time-Driven Instruction

Erickson and Mohatt (1982) investigated the interaction patterns of two expe-rienced first-grade teachers, one Indian and one Anglo, in Odawa and Ojibwa classrooms in northern Ontario. A significant contribution of this study was a comparison of the time allocated for instructional activities in the classrooms. The study documented that the Indian teacher—

- spent more time waiting for the students to finish their work—fifteen min-utes compared to an average of five minutes spent by the Anglo teacher;
- appeared to accommodate herself more sensitively to the children's rates of beginning, doing, and finishing work;
- maintained control of the students not with overt directives but by paying close attention to the rhythms of activity and judging when the students were ready for a change.

The researchers explained this behavior as reflecting a sensitivity to cultur-ally valued collaborative behavior rather than to authoritarianism. The Indian teacher's strategies involved proceeding fairly slowly and deliberately and sharing the social control: leadership was shared by teacher and students rather than divided into separate compartments—the teacher's time and the students' time. The teacher had control of the students but achieved it by pay-ing close attention to each student's progress. In this way, the students—not a predetermined schedule or curriculum directed by the teacher—set the pace for instruction.

Lipka (1991) reported similar findings in the study described earlier. He explained that Indian students have "rights and responsibilities" that are dif-ferent from those we would find in a mainstream classroom. They begin and finish a task at their own pace, within the confines of school. The teacher does not say, "OK, it's three o'clock, and it's time to leave; everybody, hand in your work." The task and the involvement of the children in the task determined the time allowed, as Gilliland explains in *Teaching the Native American* (1992, pp. 32–33):

The Native American characteristic which is probably most misunderstood is their concept of time. To European-Americans, time is very important. It must be used to the fullest. Hurry is the by-word. Get things done. They feel guilty if they are idle. They say, "Time flies." To the Mexican, "time walks." However, the Indian tells me, "Time is with us." Life should be easy going, with little pressure. There is no need to watch clocks. In fact, many Indian languages have no word for time. Things should be done when they need to be done. Exactness of time is of little importance. When an activity should be done is better determined by when the thing that precedes it is completed or when circumstances are right than by what the clock says.

The constructivist process, like the Native American instruction described above, must not be constrained by a rigid time-driven and grade-specific mastery of objectives. Constructivism views time as a way of coming to know one's world (Brooks and Brooks 1993), and coming to know one's world is a lifelong process. One's individual experiences generate understanding and the development of new knowledge; rich experiencing takes time.

REFLECTIONS ON DIFFERING PEDAGOGIES

Reflecting on the studies reported in this article leads us to conclude that learning environments in which traditional Native American ways of teaching and learning are applied share beliefs and perceptions as well as ways of evaluating and acting with environments in which constructivist-based teaching methods are applied. Therefore, it is possible to propose that a constructivist approach to teaching promises to be responsive to cultures that value Native American ways of teaching. It is also possible to suggest that both Native American pedagogy and constructivist-based pedagogy conflict with the dominant culture's traditional beliefs about teaching. Table 4.1 describes the three pedagogies and identifies differences and commonalities among them.

WHY CONSTRUCTIVISM NOW?

The proposition that culturally responsive pedagogy can be practiced only in culturally sensitive environments leads one to consider the culture that generated constructivist principles. As stated earlier, constructivism is founded on theories of cognitive psychology, philosophy, and anthropology. However, theories do not generate cultural environments; rather, cultures generate and support theories. The following questions must be posed: Why were constructivist ideas developed by educators in the dominant culture? And why is constructivism generating interest in the dominant culture? A quotation from *Curriculum and Evaluation Standards for School Mathematics* (NCTM 1989, p. 3) attributes this instructional reform to concern over societal needs projected for the twenty-first century. Reflecting on this excerpt also illuminates the reasons that American society is economically stratified.

Schools, as now organized, are a product of the industrial age. In most democratic countries, common schools were created to provide most youth the training needed to become workers in fields, factories, and shops. As a result of such schooling, students also were expected to become literate enough to be informed voters. Thus, minimum competencies in reading, writing, and arithmetic were expected of all students, and more advanced academic training was reserved for the select few. These more advanced students attended the schools that were expected to educate the future cultural, academic, business, and government leaders.

Table 4.1
A Comparison of the Principles of Three Instructional Approaches

Instruction Focus Area	Dominant Culture Pedagogy	Constructivist Pedagogy	Native American Pedagogy
Role of the teacher	The teacher generally behaves in a didactic manner, disseminating information to students.	The teacher facilitates students' learning by selecting developmentally appropriate lessons. The teacher is a "guide on the side" rather than a "sage on the stage" during these lessons.	The teacher guides the student to learn age-appropriate tasks. Conversational topics are not controlled by individual speakers.
View of learner	Students are viewed as blank slates onto which information is etched by the teacher.	Students are capable of complex problem solving. Learning is a natural and motivational experience.	Each student possesses Creator-given strengths and is born a thinker with a life mission.
Curriculum	Curriculum activities rely heavily on textbooks and workbooks.	Curriculum blends content with meaningful, real-life situations. In this way, content becomes relevant and helps the learner link knowledge to many kinds of situations.	Lessons relate to real problems that will most likely confront the student.
Time	The day is partitioned into blocks of time and content coverage. "Time on task" is considered important.	Content is taught through problem solving that may take hours, days, and even weeks.	Instruction and learning are time-generous rather than time-driven. When an activity should begin is determined by when the activity that precedes it is completed.
Concept formation	Concepts are presented part-to-whole with an emphasis on basic skills.	Concepts, procedures, and intellectual processes are interrelated. In a significant sense, "the whole is greater than the sum of its parts."	All knowledge is relational, presented whole-to-part, not part-to-whole. Just as the circle produces harmony, holistic thinking promotes sense making.
Student-to-student interaction	Students primarily work alone.	Student-to-student interaction is encouraged. Interacting with classmates helps students construct knowledge, learn other ways to think about ideas, and clarify thinking.	Caretaking patterns of extended families and bonded community interactions are replicated in group learning experiences.
Assessment	The assessment of students is viewed as separate from teaching and occurs almost entirely through testing. Testing often stratifies students and promotes competition.	Decisions regarding students' achievement are made on the basis of balanced and equitable sources that authentically document performance.	Age and ability determine task appropriateness. Learning mastery is demonstrated through performance. A Creator-ordained mission determines one's role in life, and no one mission is better than another. Competition, or situating one as better than another, is discouraged.

The educational system of the industrial age does not meet the economic needs of today. New social goals for education include (1) mathematically literate workers, (2) lifelong learning, (3) opportunity for all, and (4) an informed electorate. Implicit in these goals is a school system organized to serve as an important resource for all citizens throughout their lives.

Caution is imperative when considering these statements. We suggest that in reality the emphasis is placed on economic needs, not opportunity for all. The global market economy, not humanitarian concerns, drives this reform; however, whether driven by the economy or by humanitarianism, a shift from didactic classrooms to constructivist classrooms is real. Educators have recognized that industrial-age factory models of schooling no longer serve a nation at risk of losing marketing capabilities and societal harmony. A thought-provoking consideration is that what produced for Indian people a community-building educational process based on situated problem solving is now forcing the dominant culture to return to human and humane ways of learning, currently known as *constructivism*. The traditional dominant-culture educational system, however, evolved from industrial demands, whereas constructivism responds to the human condition.

CONCLUSION

At the beginning of this article, we suggested that attention be given not only to *what* people learn but also to *how* they learn (Philips 1983). In response to this concern, we have described a Native American way of learning and discussed its compatibility with constructivism. The following comments, although inspired by very different lived experiences, reflect this compatibility.

Brooks and Brooks (1993) describe constructivist classrooms as environments in which teachers recognize and honor the human impulse to construct understanding. They explain that teachers in these environments (p. 22)—

- free students from the dreariness of fact-driven curricula and allow them to focus on large ideas;

- place in students' hands the exhilarating power to follow trails of interest, to make connections, to reformulate ideas, and to reach unique conclusions;

- share with students the important message that the world is a complex place in which multiple perspectives exist and truth is often a matter of interpretation;

- acknowledge that learning and the process of assessing learning are, at best, elusive and messy endeavors that are not easily managed.

An Oneida kindergarten teacher identified as a practitioner of Native American pedagogy expressed similar beliefs when describing her role (Hankes 1998, p. 51):

One of the things that is passed along through culture is the belief that what we do today is going to benefit or harm the next seven generations, and that is the Great Law. So we must be mindful of how we are treating the kids. Every creation is special. Every child has strengths, and I think as a teacher you need to bring out those strengths. I don't know everything. We are a body of learners together, and they teach me things. I want them to feel that they are equal to me, that they can ask me anything, that they can be participants in their own learning, that they can help decide what's going on in our classroom. You find out their strengths by getting to know what they want to learn about and facilitating what needs to be learned that way. In a way, that's a real good balance.

REFERENCES

Basso, Kevin. "To Give Up Words: Silence in Western Apache Culture." *Southwestern Journal of Anthropology* 26, no. 3 (1970): 213–30.

Brassard, Marla R., and Larry Szaraniec. "Promoting Early School Achievement in American Indian Children." *School Psychology International* 4, no. 2 (1983): 91–100.

Brooks, Jacqueline G., and Martin G. Brooks. *In Search of Understanding: The Case for Constructivist Classrooms*. Alexandria, Va.: Association for Supervision and Curriculum Development, 1993.

Butterfield, Robin A. "The Development and Use of Culturally Appropriate Curriculum for American Indian Students." *Peabody Journal of Education* 61, no. 1 (1983): 50–66.

Cahape, Patricia, and Craig B. Howley. *Indian Nations at Risk: Listening to the People.* Charleston, W. Va.: Appalachia Educational Laboratory, ERIC Clearing House on Rural Education and Small Schools, 1992.

Cazden, Courtney. *Classroom Discourse: The Language of Teaching and Learning.* Portsmouth, N.H.: Heinemann Educational Books, 1988.

Collins, Allan, John Seeley Brown, and Susan E. Newman. "Cognitive Apprenticeships: Teaching the Craft of Reading, Writing, and Mathematics." In *Knowing Learning and Instruction: Essays in Honor of Robert Glaser*, edited by Lauren B. Resnick, pp. 453–94. Hillsdale, N.J.: Lawrence Erlbaum Associates, 1989.

Erickson, Frederick, and Gerald Mohatt. "Cultural Organization of the Participation Structures in Two Classrooms of Indian Students." In *Doing the Ethnography of Schooling*, edited by George Dearborn Spindler, pp. 131–74. New York: Holt, Rinehart & Winston, 1982.

Eriks-Brophy, Alice, and Martha B. Crago. "Transforming Classroom Discourse: Forms of Evaluation in Inuit IR and IRe Routines." Paper presented at the annual meeting of the American Educational Research Association, Atlanta, Ga., April 1993.

Foster, S. "Out of School Mathematics Teaching: Content and Instruction as Reported by Bad River Anishinabeg." Master's thesis, University of Wisconsin—Madison, 1989.

Gilliland, Hap. *Teaching the Native American*, 2nd ed. Dubuque, Iowa: Kendal Hunt, 1992.

Glatthorn, Allan A. "Constructivism: Implications for Curriculum." *International Journal of Educational Reform* 3, no. 4 (1994): 449–55.

Good, Thomas L., and Jere Brophy. *Looking in Classrooms.* London: Harper & Row Publishers, 1987.

Goodenough, Ward H. *Culture, Language, and Society.* Reading, Mass.: Addison-Wesley Publishing Co., 1971.

Hallowell, A. Irving. *Culture and Experience.* Philadelphia: University of Pennsylvania Press, 1955.

Hankes, Judith Elaine. *Native American Pedagogy and Cognitive-Based Mathematics Instruction.* New York: Garland Press, 1998.

Indian Nations at Risk Task Force. *Indian Nations at Risk: An Educational Strategy for Action.* Washington, D.C.: U.S. Department of Education, 1992.

Jackson, Philip W. *The Practice of Teaching.* New York: Teachers College Press, 1986.

Leavitt, Robert M. "Storytelling as Language Curriculum." In *Actes du quartorzième Congrès d'Algonquistes*, edited by William Cowan, pp. 27–33. Ottawa: Carleton University, 1983.

Lipka, Jerry. "Toward a Culturally Based Pedagogy: A Case Study of One Yup'ik Eskimo Teacher." *Anthropology and Education Quarterly* 22, no. 3 (September 1991): 203–22.

MacDonald, Dennis. "Stuck in the Horizon: A Special Report on the Education of Native Americans." *Education Week*, 2 April 1989, pp. 1–16.

McDermott, Raymond P. "Kids Make Sense: An Ethnographic Account of the Interactional Management of Success and Failure in One First-Grade Classroom." Ph.D. diss., Stanford University, 1976.

National Council of Teachers of Mathematics (NCTM). *Curriculum and Evaluation Standards for School Mathematics.* Reston, Va.: NCTM, 1989.

———. *Professional Standards for Teaching Mathematics.* Reston, Va.: NCTM, 1991.

Philips, Susan Urmston. *The Invisible Culture: Communication in the Classroom and Community on the Warm Springs Indian Reservation.* New York: Longman, 1983.

Ross, Allen Chuck. "Brain Hemispheric Functions and the Native American." *Journal of American Indian Education* 28 (May 1982): 72–75.

Scollon, Ronald, and Suzanne B. K. Scollon. *Narrative, Literacy, and Face in Interethnic Communication.* Norwood, N.J.: Ablex Publishing Corp., 1981.

Silliman, Elaine R., and Louise C. Wilkinson. *Communicating for Learning: Classroom Observation and Collaboration.* Gaithersburg, Md.: Aspen, 1991.

Spindler, George Dearborn, and Louise S. Spindler. *Dreamers without Power: The Menomini.* New York: Holt, Rinehart & Winston, 1971.

Swisher, Karen, and Donna Deyhle. "The Styles of Learning Are Different but the Teaching Is Just the Same: Suggestions for Teachers of American Indian Youth." *Journal of American Indian Education* 28 (August 1989): 1–14.

Tafoya, Terry. "Coyote's Eyes: Native Cognition Styles." *Journal of American Indian Education* 21 (February 1982): 21–33.

Tharp, Roland G., and Lois A. Yamauchi. *Effective Instructional Conversation in Native American Classrooms.* Santa Cruz, Calif.: National Center for Research on Cultural Diversity and Second Language Learning, University of California—Santa Cruz, 1994.

Exploring American Indian and Alaskan Native Cultures and Mathematics Learning

5

Claudette Bradley

Lyn Taylor

In their lives American Indian and Alaskan Native children have many concrete experiences that lead to a cognitive understanding of mathematics principles. These experiences serve as "readiness" activities that need to be tapped in the classroom, grades K–12, to further the mathematical growth of American Indian and Alaskan Native children (Bradley 1984; Dick 1980; Lipka 1994). In this chapter we discuss two- and three-dimensional Indian artifacts and the mathematics associated with them, three psychological frameworks supporting our views, selected research with American Indians and Alaskan Natives, and the connections among these components.

TWO- AND THREE-DIMENSIONAL AMERICAN INDIAN ARTIFACTS

American Indian baskets are three-dimensional objects that can be interpreted mathematically. Baskets are paraboloids, prisms, cylinders, and spheres. Many representational patterns are placed on their surfaces. Some designs are simply for decoration, but many have meaning and are an interpretation of the maker's environment. Many patterns are combinations of geometric shapes; some are tessellations; some, if iterated, would be fractals (see fig. 5.1).

Maidu basket pattern
(a)

Karok basket pattern
(b)

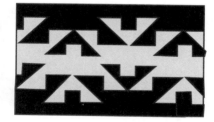

Yurok basket pattern
(c)

Fig. 5.1. Fractal (a and c) and tessellation (b) basket patterns (Source: Appleton [1971, p. 33])

Blanket and rug surfaces are two-dimensional spaces on which weavers may choose to weave a picture or a representational design. Many rugs and blankets carry a message from the culture of the maker. The patterns, such as turtle, deer, sunflower, or eagle (i.e., thunderbird) designs, are often geometric representations of our three-dimensional world. Figure 5.2 depicts several American Indian geometric designs.

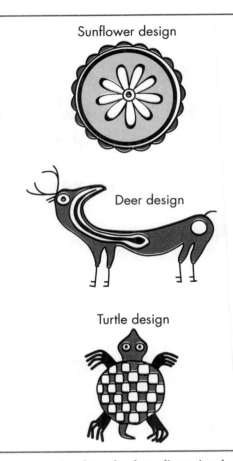

Thunderbird design, Menomini

Fig. 5.2. Two-dimensional geometric representation from the three-dimensional world (Source: Appleton [1971, pp. 8, 34])

Wampum belts are two-dimensional representations of messages, history, or treaties. They are equivalent to the Western world's use of paper. For instance, the Washington Covenant wampum belt depicts a treaty made by Washington with the Iroquois Nation (see fig. 5.3). It is a symbolic representation of people and a dwelling and has been interpreted to mean "many hearts with one mind united under a new form of government." The symmetry represents balance, order, and harmony. These attributes enhance the design's symbolism of the peoples' many hearts united under one house.

A WESTERN SOCIETAL VIEW OF TWO AND THREE DIMENSIONS

Everyone's world is three-dimensional, but schooling primarily emphasizes interacting with visual information in two dimensions, such as in books or on paper, videotapes, and computer screens. We expect that the knowledge gained from exploring two-dimensional regions will help us live in our three-dimensional world, but we do not need to understand the two-dimensional environment to survive in a three-dimensional world. Many cultures have survived without two-dimensional print and media. Survival on earth does not depend on understanding the two-dimensional world as we know it through reading books, watching videos, and playing computer games. Alan Watts (1973), a philosopher, claims that we read many miles of print to try to understand the world, when in fact, the world is multidimensional. If we look at a place in the forest, many things are happening simultaneously and we cannot possibly comprehend them in any given moment.

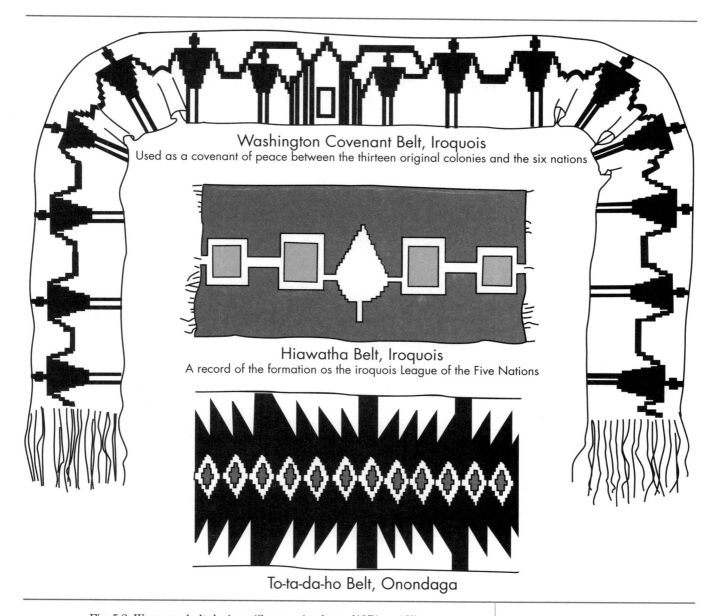

Fig. 5.3. Wampum-belt designs (Source: Appleton [1971, p. 18])

In high school mathematics we typically teach one year of plane geometry and expect three-dimensional topics to be covered as supplementary material in higher-level courses. Calculus is not the subject in which students should be struggling to understand the three-dimensional concepts involved in theorems. The development of geometric thinking should begin in preschool and continue throughout the students' grades K–12 mathematical experiences (Steen 1990).

In mathematics classrooms, objects for free play should always be accessible. The development of an awareness of space, volume, and motion should always be a part of mathematical learning in school from prekindergarten through twelfth grade (NCTM 1989). An understanding of measurement quantities and the ability to relate them to formulas will evolve naturally out of manipulations of geometric shapes and solids and the resulting rich understanding of them (Steen 1990).

Three educational psychologists support these ideas: Friedrich Froebel, the inventor of "kindergarten"; Jean Piaget; and Lev Vygotsky.

Froebel

Froebel's philosophy is that if children could be stimulated to observe geometric objects from the earliest stage of their education, these ideas would come back to them again and again during the course of their schooling, deepening with each new level of sophistication (Banchoff 1990). Students' rudimentary appreciation of shapes and forms at the nursery school level would become more refined as they developed new skills in arithmetic and measurement and later in more formal algebra and geometry (Steen 1990, p. 11).

A primitive understanding of solid geometry and plane geometry begins with the manipulation of geometric solids in the early grades. Froebel used wood, paper, and clay objects, which he presented in sequence to be manipulated and played with by children. After allotting sufficient time for students to play with objects, Froebel used a directed set of play experiences to develop mathematical concepts that children could formalize in later lessons (Banchoff 1990).

After involving students in hands-on activities with cylinders, balls, and blocks, Froebel gave them trays of tiles with mathematical patterns and directed them in play activities that involved making tile patterns. He presented students with sticks in sequences of varying lengths and had the students reposition the sticks to help them develop and recognize number patterns (Banchoff 1990).

Piaget

According to Piaget, children's learning develops in four stages. In the sensorimotor stage, the child learns to coordinate actions, which later form the basis of representational thought. From two to seven years old, the child imagines actions that he or she wants to perform during the preoperational stage. During the concrete-operational stage, the child learns through manipulating objects. In the formal-operational stage, the person uses symbols to engage in abstract thinking (Canella 1980; Piaget 1966). Piaget's work focuses on the importance of individual development.

Vygotsky

A Vygotskian perspective emphasizes the importance of play. Vygotsky believed that "creative imagination" grows out of the play of young children (Williams 1989, p. 117). He also believed that culture influences individual development. This perspective is in direct contrast with Piaget's emphasis on the progression from individual to social functioning (Taylor 1993).

Vygotsky believed that learning occurs twice: first, at the social level in collaboration with others, and second, on the individual level, where learning is internalized (Vygotsky [1934] 1978). Therefore, a rich learning environment provides a stimulating place for meaningful interactions to occur and concepts to form at the "everyday" or "spontaneous" level. "Scientific" or "theoretical" concepts are typically introduced in more-formal educational settings (Schmittau 1993, p. 30).

CONNECTING
TLINGIT
CHILDREN'S
LEARNING WITH
PSYCHOLOGY

Tlingit people are Alaskan Natives whose indigenous lands are located on the Pacific Coast of southwestern Alaska, which is adjacent to the northwestern territories of Canada. Research on the mathematics and physics skills obtained by Tlingit children while observing and participating in subsistence hunting and fishing with Tlingit adults was done by cultural anthropologist George Guilmet. Tlingits manipulate their natural environment with great skill, requiring understandings of mathematical concepts and physics. For example, they construct their wooden boats to be watertight, sturdy, and buoyant, and they build their Western style houses and fish-drying sheds with skilled craftsmanship. They customize fish nets to withstand water velocity and force, make fish lines and hooks to withstand forces of struggling fish, and dry and smoke fish for appropriate time intervals with selected humidity. They choose rifles to suit the sizes and speeds of waterfowl and terrestrial mammals, and they sharpen and file stones to maintain cutting edges on knives and axes. The Tlingits select clothing to maintain body temperature in rapidly changing environments, and they interpret the sounds made by game to determine direction and distance. These examples include only some of the subsistence activities of Tlingit adults (Guilmet 1984).

While observing and participating in such activities, Tlingit children learn. They learn to judge velocity by experiencing the velocity of the water while they are net fishing; observe the relationship between vibratory frequency and harmonics while constructing game calls; experience the relationship among the size, density, and weight of materials in constructing a deadfall; determine the flexibility of wood in relationship to its size and dryness while making snares; experience the concept of force when selecting appropriate nets, hooks, and fishing lines; master curvilinear motion when learning to aim a rifle over large distances; observe wave phenomena during travel on lakes, rivers, or ocean; and experience acceleration while traveling by boat or snowmobile (Guilmet 1984). Mathematics teachers of Tlingit children should form hands-on activities based on these relevant experiences. In this way, they can formalize the teaching of force, curvilinear motion, distance and direction, speed, velocity, acceleration, and time intervals—all concepts needed in the study of calculus.

Analyzing Guilmet's work using Piaget's model of cognitive development provides an appropriate lens for exploring the development of mathematics and physics skills in Tlingit children. Guilmet (1984) asked the question, What level of cognitive functioning do Tlingit children achieve as a result of participating in and observing subsistence activities?

Guilmet observed that Tlingit adolescents engaged in mathematics and physics tasks at the concrete-operational stage, since they "internalized actions that modify the object of knowledge, on objects at hand in their environment." Although most Tlingit adolescents are able to do mathematics and physics in the formal-operational stage, Guilmet found that they use "propositional logic and combinatorial operations" to devise scientific experiments in familiar subsistence situations (Guilmet 1984).

From a Vygotskian viewpoint, Tlingit youth use their play experiences to develop their understanding of everyday concepts. Their rich cultural environment provides the youth a stimulating place for meaningful interactions to occur. These findings with Tlingit children are likely to be applicable to learning among American Indian children.

LEARNING AMONG AMERICAN INDIAN AND ALASKAN NATIVE CHILDREN

Across American Indian and Alaskan Native cultures, children observe and participate in the activities of adults. Indian and Eskimo children playact and learn from the activities of adults engaged in life-sustaining work. The children spend hours observing adults before they indicate that they are ready to participate in and learn the same activities, which they know are important skills to learn for their adult life. Navajo children learn to weave blankets or shawls on a loom, weigh corn, grind corn, make mutton stew, classify rocks, make jewelry, trade or buy goods, play games, and care for sheep (Butler 1974).

For us to access the mathematics embedded in Indian and Eskimo activities and cultures, we need to be sensitive to the native people's view of the activities as a whole. For instance, Lipka (1994) interviewed Yup'ik elders about their use of measurement in making fish traps and story parkas. They shared information with him about the whole activity, and he has "separated" the mathematics from its context.

Indian designs have been used to create hands-on learning activities. For instance, Bradley has created Logo activites using the Sioux four-directions design (1992) and Navajo rug designs (1993). Clo Mingo's students at the Santa Fe Indian School made culturally meaningful tessellations (Taylor et al. 1991).

RESEARCH ON CLASSROOM PRACTICES

How can we transfer abilities and knowledge gained through life experiences to the mathematics classroom? We look at this question from two perspectives: first, from an understanding of the learning process of Tlingit students and how it can be replicated in classrooms, and second, from an examination of the knowledge base gained from participating in subsistence fishing, basket making, wood carving, and so on, and of appropriate ways to use that knowledge in the formal learning activities of classroom mathematics. These understandings appear to be consistent with the behavior of Indian and Eskimo children in their natural environment, where they are allowed to observe and participate in adult life-sustaining activities on their own initiative.

John E. Penick of Florida State University Developmental Research School studied the effective use of hands-on science in classrooms in the Tallahassee, Florida, public schools. He and his colleagues agreed that hands-on science was appropriate for effective learning but wanted to know the impact of conducting teacher-structured classes versus student-structured classes. In the teacher-structured (TS) classes, students manipulated objects under the direction of the teacher. The teachers in all the classes used the same lesson plans. In the student-structured (SS) classes, the students manipulated objects with no specific directions or evaluative feedback. The teachers observed the students and asked questions related to the students' work, but the students explored objects using their own ideas (Penick 1981).

Both groups used the same hands-on materials. The TS teachers continually praised and rejected the students' behavior and directed their use of the manipulatives. The SS teachers observed students without accepting or rejecting behaviors but asked nondirective questions designed to encourage cognitive development and the achievement of learning goals. Such questions can guide students in the appropriate use of the hands-on materials (Strange and Henderson 1981).

Surprising outcomes included the following: the SS students consistently exhibited on-task, lesson-related behaviors more often than the TS students did; the SS students spent more than 90 percent of their time on lesson-related activities during the ten- to twenty-week periods. Disruptive behaviors decreased significantly, and productive, on-task, lesson-related behaviors increased. Both sets of classes scored equally well on verbal creativity tests, whereas the SS students scored significantly higher in figural creativity. Hands-on learning in SS classes had the greatest impact on the low-ability students, who performed disproportionately better than the students in any other ability groups in either the TS or SS classes. This study by Penick (1981) indicates that educators need to encourage independent work without consistent direction from the teacher; this approach can potentially overcome some handicaps of initial ability. The learning process of Indian and Eskimo children similarly involves manipulating objects in a student-structured environment.

When we view learning through the lens just described, we can see that Indian and Eskimo children are engaged in mathematical thinking at the concrete-operational stage. We see subsistence activities providing concrete experiences of time, distances, velocity, acceleration, and curvilinear movement. We see interpretations of cultural experiences in the three-dimensional world being transferred onto such two-dimensional surfaces as beaded sashes, baskets, and blankets. These experiences give students opportunities to develop everyday concepts as well as provide readiness experiences for geometry, calculus, and higher mathematics.

Replicating the educational process of American Indians and Eskimos in mathematics classrooms can be a vehicle for developing formal mathematics thinking. We need to understand the knowledge base gained by American Indian and Alaskan Native students and use hands-on activities to make cognitive bridges to formal mathematics concepts.

BIBLIOGRAPHY

Appleton, LeRoy H. *American Indian Design and Decoration.* New York: Dover Publications, 1971.

Banchoff, Thomas F. "Dimension." In *On the Shoulders of Giants: New Approaches to Numeracy,* edited by Lynn Arthur Steen, pp. 11–59. Washington, D.C.: National Academy Press, 1990.

Bradley, Claudette. "Issues in Mathematics Education for Native Americans and Directions for Research." *Journal for Research in Mathematics Education* 15 (March 1984): 96–106.

———. "The Four Directions Indian Beadwork Design with Logo." *Arithmetic Teacher* 39 (May 1992): 46–49.

———. "Making a Navajo Blanket Design with Logo." *Arithmetic Teacher* 40 (May 1993): 520–23.

Butler, Katie B. *Walking in Navajo Footprints.* Houston, Tex.: Houston University, College of Education, 1974.

Cannella, Gaile S. "Cognition: Development and Education." Cedar Falls, Iowa: University of Northern Iowa, 1980.

Dick, Alan. *Village Science: A Resource Handbook for Rural Alaskan Teachers.* McGrath, Alaska: Iditarod Area School District, 1980.

Fletcher, J. D. *What Problems Do American Indians Have with Mathematics?* Provo, Utah: WICAT Education Institution, 1983.

Froebel, Friedrich. *Education by Development.* New York: D. Appleton & Co., 1899.

Guilmet, George M. "American Indian and Alaska Native Education for High Technology: A Research Strategy for Creating Culturally Based Physical Science and Mathematics Education." Paper presented at the annual meeting of the Society for Applied Anthropology, University of Puget Sound, Tacoma, Washington, March 1984.

Lipka, Jerry. "Culturally Negotiated Schooling: Toward a Yup'ik Mathematics." *Journal of American Indian Education* (spring 1994): 14–30.

National Council of Teachers of Mathematics (NCTM). *Curriculum and Evaluation Standards for School Mathematics*. Reston, Va.: NCTM, 1989.

Penick, John E. "Teacher Behavior Does Make a Difference in Hands-On Science Classrooms." *School Science and Mathematics* 81 (May-June 1981): 412–22.

Piaget, Jean. *The Psychology of Intelligence*. Totowa, N.J.: Littlefield, Adams & Co., 1966.

Schmittau, Jean. "Vygotskian Scientific Concepts: Implications for Mathematics Education." *Focus on Learning Problems in Mathematics* 15 (spring and summer 1993): 29–39.

Strange, Johanna, and Stephen A. Henderson. "Early Adolescence: Classroom Management." *Science and Children* 19 (November-December 1981): 46–47.

Steen, Lynn Arthur, ed. *On the Shoulders of Giants: New Approaches to Numeracy*. Washington, D.C.: National Academy Press, 1990.

Taylor, Lyn. "Vygotskian Influences in Mathematics Education, with Particular Reference to Attitude Development." *Focus on Learning Problems in Mathematics* 15 (spring and summer 1993): 3–17.

Taylor, Lyn, Ellen Stevens, John Peregoy, and Barbara Bath. "American Indians, Mathematical Attitudes, and the Standards." *Arithmetic Teacher* 38 (February 1991): 14–21.

Vygotsky, Lev. *Mind in Society*. 1934. Reprint, Cambridge, Mass.: Harvard University Press, 1978.

Williams, M. "Vygotsky's Social Theory of Mind." *Harvard Educational Review* 59 (1989): 108–26.

The Learning of Geometry by the Inuit

A Problem of Mathematical Acculturation

6

> Learning and thinking are activities that are always situated within a cultural context.
>
> —Jerome Bruner

In a previous study, three of the authors examined the influence of children's spatial environments on their development of spatial skills in a situation focusing on the manipulation of small objects and their two-dimensional representations (Pallascio et al. 1990). These observations suggested that Inuit children and children from an urban environment, inhabiting as they do dissimilar spatial environments, differed from each other in their perception and representation of the geometric properties of different objects. They differed as well in spatial skills. For example, children in the control group performed better than Inuit children in identifying topologically identical forms, whereas the Inuit children performed better than the children in the control group in generating forms by truncation (e.g., a hexahedron from a rectangular pyramid). We were able to show that the spatial environment influences the development of spatial relationships. Results such as these have prompted us to consider the cultural context in which spatial skills develop.

Studies on contextualization have indicated the desirability of considering an individual's cultural environment. According to Bruner (1996, p. 7),

> All mental activity is culturally situated. Indeed, it is impossible to understand mental activity if one fails to account for the cultural environment and the resources it makes available—in other words, the myriad details which shape the mind and determine its scope. Learning, remembering, speaking, imagining—all of this becomes possible only because we participate in a culture.

In keeping with this outlook, we have previously proposed a series of contextualized activities involving the generation of three-dimensional forms (Pallascio, Allaire, and Mongeau 1992). In addition, we subscribe to the notion that "any educational practice which proposes to further strengthen the mind must make of 'thinking out the act of thinking' a centerpiece of its action" (Bruner 1996, p. 36). For that reason, we view an improvement in our understanding of the representations, metacognitive processes, and affective reactions related to spatial skills as central to developing contextualized activities in a way that will prove useful to mathematical and professional education.

In this article, we take up a number of related issues, the most notable of which are the education of young Inuits in spatial and geometric skills, the

Richard Pallascio
Richard Allaire
Louise Lafortune
Pierre Mongeau
with
Justin Laquerre

The research and analysis reported in this chapter was supported by the Social Scientand Humanities Research Council of Canada (SSHRC) under grant number 884-94-0006. The opinions expressed do not necessarily reflect the views of SSHRC.

phenomenon of mathematical acculturation, the objectives of the research project thus conducted, the methodology employed, a synthesis of the observations made, and the main recommendations stemming from the entire research process.

THE EDUCATION OF INUIT YOUTH

Until very recently, for a range of sociohistorical reasons, many Inuit youth were forced to complete their professional education in institutions in a foreign culture. This circumstance presented a number of problems, particularly as far as their training in mathematics was concerned. A wide gap separated their expectations, on the one hand, from the reality of their school experiences, on the other. It should also be pointed out that the teaching methods now being used in northern Quebec (a new professional training center has recently opened in Nunavik) are the same as those employed in southern Quebec. Even if Inuit children have developed a solid command of spatial perception and an exceptional visual memory (Pelley 1991), they do not perceive the same way that other children do (Pallascio, Allaire, and Mongeau 1993), and consequently, they continue to have difficulty in geometry and, later, in various training programs (e.g., piloting programs, programs in the building trades).

INUIT SPATIAL AND GEOMETRIC SKILLS

In all cultures, humans develop the capacity to recognize forms and also to transform them (Bishop 1991; Lean and Clements 1981). The spatiogeometric operations associated with these skills are of various kinds: estimating distance reliably, identifying the movements in a system of cogs and gears, locating a point in a three-dimensional space, identifying segments that are fitted together after being moved, recognizing objects that have been rotated in space, and so on. These spatial skills are of significant use in the Inuits' environment (e.g., in locating oneself, heading in a particular direction, going from one place to another).

With an awareness that spatial skills are influenced by cultural context and spatial environment (Vandenberg and Hakstian 1978; Pallascio, Allaire, and Mongeau 1993), a number of researchers have developed a frame of reference to account for cultural differences in spatial representation (Pinxten, van Dooren, and Harvey 1983). According to this framework, there are three different types of space: physical, sociogeographic, and cosmological. We are concerned with notions involving sociogeographical space—for example, proximity, dimension, reduction according to scale, approximation and calculation of distances, parallelism, and cardinal points.

THE PHENOMENON OF MATHEMATICAL ACCULTURATION

Mathematical acculturation refers to the process by which a social group, and ultimately each of its members, actively constructs mathematical knowledge on the basis of experience in a sociocultural environment that is not their own. Ethnomathematical studies have shown that this process of acculturation often leads to intellectual impasses. The example of Brazilian children working as street vendors is an outstanding case in point. At a very early age, the children studied developed original, effective algorithms for sales operations requiring the counting of money (selling, returning change, etc.), and they earned passing grades in 98 percent of the exercises they were given that referred to this context. However, their performance dropped dramatically—to 37 percent—in exercises that were formally identical but referred to a school context (Carraher, Carraher, and Schlieman 1985). Similarly, Inuit students, despite having developed mathematical concepts in context (Pelley 1991), had to struggle to be able to assimilate concepts derived from the geometry of others. Although the Inuit continue to be geographically isolated, they are in contact with the rest of the world (by means of computers and television broadcasts by satellite, etc.). As a result, they are confronted with the process of acculturation. On the one hand, they wish to preserve their traditions and ways of doing things, but on the other hand, they also wish to participate in the "foreign" world in which they find themselves, by developing a number of skills associated with that world, such as piloting an airplane or using technologically advanced instruments to navigate at sea.

A major question comes to mind: Is it possible in the long term to convert this process of acculturation into a process of enculturation, which makes greater allowances for the student's culture and can involve, for example, a reformulation of the "didactic contract" (i.e., the pupils' implied expectations of their teacher, and vice versa) (Brousseau and Centeno 1991) and an ethnomathematical interpretation of the knowledge that is to be acquired? Our reply to this question can only be incomplete, of course. Nevertheless, our hope is that we shall be able to bring out a number of elements on the basis of which a response might be ventured. But before proceeding any further, we should like to define contextually the frame of reference that guided this research project.

AN ETHNO-MATHEMATICS RESEARCH PROJECT

Ethnomathematics is a "cultural anthropology of mathematics and mathematics teaching" (Gerdes 1995). This area of research involves studying the relationships between the culture of a people and mathematics—in this instance, geometry—and the ways in which mathematics is actualized within the culture. Mathematics is something all societies do, although each society constructs its own representations, just as every individual constructs his or her own representations. The choice of representation is also affected by technological developments from one generation to another, as, for example, in the shift from the slide rule to the display calculator.

Background

The type of mathematics discussed here is primarily school mathematics. Inuit culture refers to the customs of the inhabitants of northern Quebec, including their attitudes and ways of living and interacting, in addition to their native language and their life experiences, whether these are related to Inuit traditions or to new lifestyles that have sprung from contact with other people and other cultures. "A culture is a set of people who have a set of shared experiences" (Amir and Williams 1994). A major cultural fact worth pointing out is that although the Inuit possess their own language, the language of instruction beginning in grade 3 is French or English, except in classes about the Inuktitut language or Inuit culture.

According to the constructivist theory of the development of knowledge (von Glasersfeld 1995), empirical knowledge serves as the basis of all learning (Angers and Bouchard 1986). All learning is necessarily cultural in character. This view is reinforced by socioconstructivist research, which has demonstrated the importance of social interactions and communication in the construction of learning experiences (Cobb and Bauersfeld 1995). Pinxten (1994) has even asserted that learning is a cultural phenomenon and that the contents of the learning process are culturally specific.

Phases of Enculturation

On the basis of research by McIntosh (1983) and Leder (1995), five progressive phases of enculturation can be identifed. They do not exist in the absolute; components of these different phases may coexist within one and the same community.

Phase 1: Mathematical acculturation

All mathematical ideas are influenced by the culture of those who have constructed them. In the context of acculturation, the presentation of new notions is achieved by means of placing these notions in opposition to the cultural influences of the people who must assimilate them.

Phase 2: Mathematics that includes cultural connotations

In the second phase of enculturation, the presentation of new notions is achieved by using traditional objects and terms having a cultural connotation. For example, symmetry might be demonstrated using the symmetrical characters for syllables in the Inuktitut alphabet, such as "pa" (<) and "poo" (>).

Phase 3: A cultural split

In this phase, members of a culture adopt two different perspectives on mathematics, since in a certain way the individuals straddle two cultures—their traditional culture and the elements of other cultures that are developing in the midst of their own society. In this context, how do children view mathematics? It appears that they have two frames of reference, one developed in connection with school mathematics and another related to the problems of everyday life. A dissociation may occur between these two frames. On the one hand, students may learn school mathematics so as to pass their examinations, to acquire a job of their choice, or to enjoy various social advantages. On the other hand, they may learn mathematics of a type that is useful to them in everyday situations or that is intellectually amusing.

Phase 4: Cultural interactions

At this phase of enculturation, we are at the heart of the relationships among mathematics, knowledge that children need to learn, world views, and culture. The traditional categories encountered in school mathematics are culturally colored; alternative approaches might prove more appropriate. For example, categories based on such applications as counting, locating, measuring, drawing, playing, and proving (Bishop 1988) are all examples of categories that mathematically acculturated societies can readily assimilate. In addition, cognition-in-action (Brown, Collins, and Duguid 1989) and experience-based learning (Boud, Cohen, and Walker 1993) bring out the usefulness of contextualized mathematics and of activities that offer some social justification for learning mathematics.

Phase 5: Mathematical enculturation

This phase consists of developing alternative approaches to mathematics teaching that are capable of encompassing—

- aspects of the traditional culture (world views, particular knowledge, specific attitudes toward learning, etc.);
- aspects of the popular culture (expressions, behavior, students' experiences, etc.);

- mathematics (contents, processes, teaching methods) related to the educational objectives of the community;
- the needs of students (social, economic, etc.);
- the integration of learning experiences in a way that is mindful of community needs.

Mathematical acculturation among the Inuit creates a major psychological gap between indigenous social practices that offer a potential for association with spatial and geometric skills and institutionalized knowledge such as that found in the mathematics textbooks used in classrooms. For example, the Inuit are capable of determining the distance to the sea by smelling the air for traces of salinity. This "sensory gauge" is fundamentally foreign to the culture of school mathematics. In school, then, Inuit youth fail to solve certain types of problems because, from their viewpoint, they "have no way of gaining access to the mysticism of whites" (a quote from a young Inuit), which they assume would allow them to solve these problems! In other words, since the situations are based on different cultural representations, the difficulty consists in connecting the experience of the subjects themselves and the conceptions that these subjects have of the world around them.

Objectives of Our Research

The general hypothesis underlying our approach is that mathematical acculturation is influenced overall by students' metacognitive activity in relation to their levels of interest and motivation in the learning situations in which they find themselves (see fig. 6.1). Accordingly, the main objectives of this educational research project were—

1. to become familiar with the most important and the most affectively meaningful sociogeographic spatial representations in the daily, contextualized experience of Inuit youths;
2. to understand more fully a number of determinants in the metacognitive processes of Inuit students when they are faced with solving problems related to spatial ideas;
3. to identify the sociocultural elements that teachers and curriculum developers should consider along with students' interest and motivation when planning instruction;
4. to design and test a number of spatial problems to be solved, in order to identify the various aspects that teachers and curriculum developers should consider when designing curricula specifically for Inuit students and to field-test the resulting ideas.

Methodology

As we developed our research instruments and performed experiments, we attempted to take into consideration the fact that the first language of the subjects is not French, the researchers' language, and that their culture also differs from that of the researchers. The project took place during two phases. However, we validated our learning activities in a preexperimental phase with older Inuit students living in southern Quebec.

The first experimental phase

The first experimental phase took place in the northern Inuit village of Puvirnituq in 1995 with twelve Inuit students (eight girls and four boys) in

TOWARD A CULTURALLY CONSTRUCTED MATHEMATICS

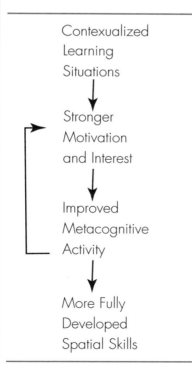

Contexualized Learning Situations

↓

Stronger Motivation and Interest

↓

Improved Metacognitive Activity

↓

More Fully Developed Spatial Skills

Fig. 6.1. A schematic of the authors' hypothesis

French-speaking classes in secondary school grades 3 to 5 (ages fifteen to twenty-one). We collected our data using tape recordings made when the subjects were working on five tasks. We also collected data during a second taping session after the researchers and the interpreter had examined the first tape and attempted to understand what the subjects had said and accomplished. We then analyzed these tape recordings using three grids that took into account cognition and metacognition in addition to the students' interest in the content of the five learning activities. One of these activities, for example, asked the participating subjects (three to five students) to draw a map of the village, indicating the position of a number of main buildings (school, hospital, cooperative, post office) as well as their own homes. No limitations were placed on the way students went about drawing this map. All necessary materials were made available (large sheets of paper, compasses, rulers, crayons, etc.). Once these village maps were finished, they were hung on the classroom wall, and each subject explained his or her drawing. Students answered a number of questions that required them to make clear the procedure they had used to draw the maps.

The second experimental phase

The second experimental phase took place in Puvirnituq during 1995–1996. We met fourteen students (seven boys and seven girls), ages fifteen to nineteen (three boys and four girls in secondary school grade 3, one boy and one girl in secondary school grade 4, and three boys and two girls in secondary school grade 5) during mathematics class over a ten-day period. A test focusing on metacognition was administered during the French class to leave more time for carrying out projects and to elicit explanations in French of certain test items. Data were collected using questionnaires that dealt with cognitive and metacognitive aspects of the tasks and with the interests of Inuit students. Two members of the research team also kept a journal in which they recorded their observations while these youths carried out their projects. We also fleshed out these data by means of individual interviews with several students.

By using project-oriented tasks during the first experimental phase, we were able to collect a certain amount of information as students solved problems requiring geometric and spatial skills. However, we noticed Inuit students' enthusiasm for certain types of activities. Hence, we wished to expand the possibilities for observation by proposing a number of tasks that would be of interest to the Inuit students and that would foster a certain autocontextualization. We thus developed a series of projects that teachers proposed to the Inuit youngsters approximately two months before we arrived. From the seven choices offered, teams of students each selected the one that most interested the team's members. Each proposed project could involve creation or reproduction. The students selected six of the seven projects: (1) building a model of an Inuit home, (2) creating spatial forms, (3) inventing spatial and geometric games, (4) creating a school bag, (5) creating a three-dimensional puzzle, and (6) building a model of an Inuit youth center. For the last project, one secondary school grade 5 team (two boys and one girl) chose to produce a scale model of an Inuit youth center. To learn how to make a design, the students first had to make a scale drawing of an existing house and produce a plan for their own house and then build their model. Each team or individual presented its project to the other people at the school and had the opportunity to explain the process involved in their project.

Results and Analysis of the First Experimental Phase

Observations relating to spatial competencies (first objective)

We intended the first experimental phase to be exploratory. This phase enabled us to confirm that the proposed activities brought into play the expected concepts and skills. In addition, it confirmed the presence of a number of weaknesses that had been observed in students' geometric and spatial concepts during a preexperimental phase among Inuit students living in southern Quebec. At this stage, we were also able to clarify a number of linguistic difficulties that had been noted in the course of the preexperimental phase. An analysis of verbatim transcripts and students' written (or other) productions showed that the intellectual operations involved in the shift from two-dimensional representations to three-dimensional representations (and vice versa) were particularly difficult for these students.

Observations relating to metacognitive processes (second objective)

In activities in which students worked in teams and cooperated extensively with one another, as in preparing a village map, the metacognitive aspects of a number of student interactions were apparent. In more solitary types of exercises, however, the students remained quiet and appeared to pay little attention to what others were doing; in such instances, gaining access to their metacognitive processes was difficult. Moreover, verbatim transcripts of these solitary exercises showed that metacognitive features of such situations were the direct outcome of interventions by the researchers, who used many questions to help the pupils express their mental processes. Thus, although it cannot be asserted that a group task, such as creating a map of the village, results in greater metacognition, it certainly does make for easier observation of metacognition. In addition, group discussion with pupils on the processes that they use to do such tasks appears to contribute to the metacognition of the students. Thus, on analyzing the data collected during this phase of experimentation, the research team decided that the protocol of the final experimental phase would allow the researchers to intervene more openly to encourage students to express their cognitive processes and make them explicit.

Furthermore, we noticed that activities involving producing something by means of a visual support that was kept continuously on display before students, such as the village map, elicited "regulation" interactions—that is, interactions designed to correct a production in progress by pointing to desirable modifications in strategies. Listening activities, however, led to discussions about planning and design.

Observations about the sociocultural elements related to interest and motivation (third objective)

The tasks used during the first phase of experimentation did not elicit much enthusiasm. The students manifested the weakest interest in the most decontextualized activities—that is, the activities having the fewest cultural connotations, whereas the more contextualized activities, which had greater cultural or everyday connotations, drew a more positive response.

The presence of the classroom mathematics teacher greatly encouraged the Inuit youths to take part in the learning activites. The teacher was well known and had been in the Inuit community for more than sixteen years, and he was able to make clear to the students what the community stood to gain from this project—for instance, new perspectives for technical training. Also, supporting observations made during the preexperimental phase, the process of Inuit mathematical acculturation in spatial skills appeared to be influenced overall by problems of language. These difficulties included a lack of, or ignorance of, precise words in Inuktitut; the occasionally difficult shift from Inuktitut to French; the omnipresence of English in their everyday lives; and so on.

This first experimental phase led our team of researchers to the conclusion that the process of acculturation in northern Quebec is primarily school-related. Establishing relationships between the everyday lives of students and problem situations remains a source of ongoing difficulty. The concepts that are learned at school appear to remain inside the classroom.

Observations relating to aspects of teaching (fourth objective)

The five tasks that were developed during the first phase of experimentation could be placed along a continuum extending from the most contextualized to the least contextualized. In addition to the making of a village map, the contextualized activities consisted of (*a*) using diagrams to make knots that the Inuit are familiar with and (*b*) devising a two-dimensional drawing to make a scale model of a box that would be hooked onto the back of a snowmobile. Decontextualized activities consisted of (*a*) following verbal instructions to reproduce a spatial form made up of a number of stacks of small cubes, with the students giving the instructions hidden from the students constructing the forms, and (*b*) enlarging drawings of varying complexity according to the instruction "four times bigger." The continuum of contextualization was relative to a number of visual and "geographical" elements—for example, using the village to make the map and using a sleigh to make a model. Unfortunately, the type of contextualization resulting from this type of task did not appear to be connected to the utilitarian and functional aspects of the everyday lives of these Inuit students.

Results and Analysis of the Second Experimental Phase

Observations relating to spatial notions (first objective)

In the second phase of experimentation, test results showed a positive correlation between cognition and interest. From this correlation, it may be seen that the Inuit youths had greater mastery of what interested them more or that they took a greater interest in what they had more mastery of. This observation partially corroborated our initial hypothesis that in the acquisition of geometry skills, the process of mathematics enculturation would stand to gain from contextualized situations that awakened students' interest. After observing difficulties in the shift from two dimensions to three during the first experimental phase, we integrated into the second phase of our experimentation a number of tasks involving transposition, such as the use of projections or three-dimensional fold-out figures. Performing these tasks did indeed trigger students' learning of geometric and spatial concepts and thus fostered mathematical enculturation. As a result of having been integrated into activities capable of stimulating the interest of students, these transpositions generated situations that resulted in positive learning experiences in geometry.

Observations relating to metacognitive processes (second objective)

Supporting the observations and suggestions made during the first phase of experimentation, greater interaction between the students and the team of researchers effectively stimulated metacognitive reflection in the participating students during the second phase of experimentation. To be more precise, the interventions of the professors and researchers generally awakened students to elements to consider in solving a problem or a situation. In addition, accomplishing projects with peers also stimulated the students' metacognitive development. For example, interaction among students on a team project gave rise to the development of metacognition about the strategies and procedures to be used in solving spatial problems or allowed students to identify their personal resources. In short, teamwork made it possible for these students to develop metacognitive skills related to insight, planning, and regulation. Students observed one another as they solved problems and served as models for one another. Furthermore, it is worth noting that all the teams except for one freely and spontaneously formed themselves from students having unequal but complementary strengths in metacognition. These teams included at least one student who had high scores and one student who had low scores on the test of metacognitive activity.

Observations about the sociocultural elements related to Iinterest and motivation (third objective)

The project's instructional approach was adopted during the second experimental phase. The researchers allowed students to choose among several activities that theoretically would allow them to develop geometric and spatial knowledge and skills. This approach made it possible to consider and stimulate students' interest in geometric and spatial ideas and to circumvent the problems inherent in white researchers' from southern Quebec devising contextualized tasks for members of the Inuit culture.

This approach was successful, stimulating and maintaining a significant level of interest among students. All the projects, with the exception of the sculpture project, which was not chosen by any team of Inuit pupils, generated comparable and continually growing interest. In addition, the students showed no small degree of pride when they presented their productions during an exhibit organized at the end of the experiment. The prospect of these presentations helped bolster their desire to pursue their projects to completion. To some degree, the exhibit provided meaning and a goal for the process.

The results of a questionnaire used in evaluating the experiment bore out the preceding observations. Most of the students enjoyed learning through projects and indicated that they would like to do another one. They particularly liked the active aspect of this type of learning. Furthermore, they liked making a public presentation of their work and declared that they were very satisfied with their project and their teamwork. They believed that the projects had helped them in mathematics and had allowed them to learn something new. Moreover, they said that if they had the opportunity to start over, they would do a new project, different from the first one. In response to an open question that asked what portions of the experiment they liked and disliked the most, they explicitly stated that they did not like answering questions, either in a questionnaire or during an interview. However, they thought it was important to keep a journal.

Observations relating to teaching (fourth objective)

The interest shown in the projects appeared to be related to the metacognitive development of students. The scores obtained using quantitative instruments showed a very strong correlation ($r = .82$) with students' stated overall interest in the proposed activities as a whole. This correlation carried over to activities involving design and those involving reproduction. However, no significant difference appeared in the correlations with the interest generated by these two types of activities.

DISCUSSION AND RECOMMEN-DATIONS

Notwithstanding the relatively universal character of mathematical concepts and the symbolism by which human beings are able to use them effectively, every community develops its own mathematical instruments. Ultimately, every generation has to reappropriate these concepts using the new technologies that it has available (e.g., calculators, computers), just as every individual must construct his or her own mathematical knowledge and combine it with institutionalized mathematics of the type that is taught in school programs.

On the basis of these considerations, the results of this research project prompt us to offer several recommendations to teachers:

- *Recommendation 1:*
  Encourage the implementation of a project-based teaching approach.

- *Recommendation 2:*
  Propose projects that require the use of geometric transpositions.

- *Recommendation 3:*
  Let students choose their own partners.

- *Recommendation 4:*
  Intervene frequently for the purpose of making metacognitive activity explicit.

- *Recommendation 5:*
  Propose projects that have a relationship to high school mathematics classes and training in certain professional programs.

- *Recommendation 6:*
  Develop strategies of enculturation as a means to further the development of solid preprofessional training in spatial and geometric skills.

The implementation of a project-based teaching approach makes it possible for learning experiences to be autocontextualized. By the same token, it fosters metacognitive activity (such as gaining insight into personal learning processes), facilitates the teacher's planning of learning activities (such as solving mathematical problems), and aids regulation, thus making it possible to adjust for any obstacles to cognition that arise.

The use of activities that draw on transpositions (e.g., going from a geometric drawing to a model, and vice versa) should be encouraged because (*a*) spatial and geometric skills are required in numerous professional training programs, such as those for technical drawing, pattern making, industrial mechanics, and building maintenance; (*b*) the cultural patterns of the Inuit tend to favor skills in determination and generation (spatial visualization) over skills in structuration and classification (spatial relationships); and (*c*) transposition skills are central to transferring skills between these two areas.

The young Inuit whom we met liked to work in teams and had natural cultural aptitudes that were based on cooperation. We noticed that students whose metacognitive skills complemented one another's formed teams on their own, without the intervention of the teacher.

Increasingly, metacognitive behavior is viewed as playing a central role in all meaningful cognitive development. In addition, the Inuit high school students who participated in this research were receptive to the requests and encouragement of the researchers to discuss their metacognition throughout the entire research project. For both these reasons, teachers should actively stimulate metacognitive activity, primarily in an inductive fashion (e.g., through discussions of the learning objectives, the problem-solving process, the knowledge and skill involved, the corrective measures that need to be taken with respect to the objectives of a project).

Full enculturation requires a phase of contextualization of mathematics learning experiences. This phase should be allowed to unfold in accordance with the aspirations of a community for its young people. So that geometric and spatial learning experiences in the high school program become genuinely meaningful, it would therefore be appropriate to show how such learning experiences will be useful at some future time in various professional applications, such as in reading plans or producing technical drawings.

A number of professions that appear to be a priority for the development of the Inuit nation require solid skills in spatial perception and reception. These skills also appear to have a solid cultural basis. For both these reasons, these skills should be made one of the priorities of general education programs.

Teachers might use these six recommendations as parameters in planning projects that foster mathematical enculturation. Figure 6.2 shows an example of a project and its component steps, which teachers might regard as a model for integrating the preceding recommendations into their interventions.

Project (*recommendation 1*): Develop proposals for one or more permanent three-dimensional play structures for the local primary school playground.

School level: High school (fourteen- to seventeen-year-olds)

At various steps, instructions are provided by the teacher to facilitate students' metacognitive reflection (recommendation 4) (personal stock taking, planning, evaluation and correction while the project is being developed, etc.)

Step 1: Documentation and researching the literature (documents, Internet, etc.)

Step 2: Brainstorming with the entire class to find ideas for play structures (geodesic domes, labyrinths, etc.)

Step 3: Group selection of a few proposals for structures

Step 4: Self-teaching in teams of four students according to students' interest in each of the structures selected (*recommendation 3*)

Step 5: Teaching by teachers of a number of general notions (visualization or analog drawing, proxemics, scale, spatial limitations, perspective and projections, integration of art and mathematics, etc.)

Step 6: Scale drawings (*recommendation 5*) of views (section, elevation, and plan) of each team's structure

Step 7: Inclusion of local cultural features (*recommendation 6*) in structures (e.g., truncated pyramids that recall the form of the blocks of snow used to construct igloos)

Step 8: Construction of 1:20 scale models of proposed structures (*recommendation 2*)

Step 9: Presentation of results to the primary school parents' committee, which selects one or more proposals for actual construction

Fig. 6.2. The steps in a sample project

We have characterized the cognitive, metacognitive, and motivational types of development according to the phases derived from our theoretical framework. To aid in this task, we have drawn on the observations and analyses conducted over the course of this research project, according to the dimensions of cognition, metacognition, and interest (see table 6.1).

Table 6.1
Cognition, Metacognition, and Interest according to the Theoretical Phases of Encultration

	Cognition	Metacognition	Interest
Mathematical acculturation	Mathematical concepts offering split meanings	Unfacilitated metacognitive development	Opposing cultural influences
Cultural connotations	Use of traditional objects	Cultural evocations capable of triggering a self-questioning process	Cultural references
Cultural split	School mathematics dissociated from everyday mathematics	Two frames of mind coexist	Extrinsic objectives versus intrinsic objectives
Cultural interactions	Mathematics applied to everyday situations	Subject's activity based on his or her experience	Interactions among people, their culture, and mathematics
Mathematical enculturation	Culturally reconstructed or locally developed mathematics	Reasoned and responsible knowledge	Alignment of learning experiences with the needs and knowledge of the community

FROM MATHEMATICAL ACCULTURATION TO MATHEMATICAL ENCULTURATION

By means of this characterization, it is possible to determine the degree of enculturation typifying a given sector of activity in a community. The table also makes it possible to target the "didactic contracts" and the teaching approaches that are best suited to making a gradual transition from acculturation to a form of mathematical enculturation.

REFERENCES

Amir, Gilead, and Julian Williams. "The Influences of Children's Culture on Their Probabilistic Thinking." In *Proceedings of the 18th International Conference for the Psychology of Mathematics Education (PME), Lisbon, Portugal*, edited by Joao Pedrode Ponte and Joao Phillipe Matos, vol. 2, pp. 24–31. ERIC document #ED38357. Lisbon, Portugal: PME, 1994.

Angers, Pierre, and Colette Bouchard. *L'appropriation de soi*. Montréal, Quebec: Bellarmin, 1986.

Bishop, Alan J. "Mathematics Education in Its Cultural Context." *Educational Studies in Mathematics* 19, no. 2 (1988): 179–91.

———. *Mathematical Enculturation*. Dordrecht, Netherlands: Kluwer Academic Publishers, 1991.

Boud, David, Ruth Cohen, and David Walker. "Introduction: Understanding Learning from Experience." In *Using Experience for Learning*, edited by David Boud, Ruth Cohen, and David Walker, pp. 1–17. Buckingham, U.K.: Open University Press, 1993.

Brousseau, Guy, and Julia Centeno. "Rôle de la mémoire didactique de l'enseignant: Recherches en didactique des mathématiques." *La Penseé sauvage* 11, no. 2–3 (1991): 167–210.

Brown, John S., Allan Collins, and Paul Duguid. "Situated Cognition and the Culture of Learning." *Educational Researcher* 18, no. 1 (1989): 32–42.

Bruner, Jerome. *L'éducation, entreé dans la culture*. Paris, France: Retz, 1996.

Carraher, Terezinha, David W. Carraher, and Analúcia Dias Schlieman. "Mathematics in the Streets and in Schools." *British Journal of Developmental Psychology* 3, no.1 (1985): 21–29.

Cobb, Paul, and Heinrich Bauersfeld. *The Emergence of Mathematical Meaning: Interaction in Classroom Cultures*. Hillsdale, N.J.: Lawrence Erlbaum Associates, 1995.

Gerdes, Paulus. "L'ethnomathématique en Afrique." *P.L.O.T.* 70 (1995): 21–25.

Lean, Glen A., and Ken A. Clements. "Spatial Ability, Visual Imagery, and Mathematical Performance." *Educational Studies in Mathematics* 12 (1981): 267–99.

Leder, Gila C. "Learning Mathematics: The Importance of (Social) Context." *The New Zealand Mathematics Magazine* 32, no. 3 (1995): 27–40.

McIntosh, Peggy. *Phase Theory of Curriculum Reform*. Wellesley, Mass.: Center for Research on Women, 1983.

Pallascio, Richard, Richard Allaire, Laurent Talbot, and Pierre Mongeau. "L'incidence de l'environnement sur la perception et la représentation d'objets géométriques." *Revue des Sciences de l'Éducation* 16, no.1 (1990): 77–90.

Pallascio, Richard, Richard Allaire, and Pierre Mongeau. "Spatial Representation and the Teaching of Geometry." *Structural Topology* 19 (1992): 71–82.

———. "Spatial Representation of Geometrical Objects: A North-South Comparison." *Inuit Studies* 17, no. 2 (1993): 113–25.

Pelley, David F. "How Inuit Find Their Way in the Trackless Arctic." *Canadian Geography* (August-September 1991): 58–64.

Pinxten, Rik. "Anthropology in the Mathematics Classroom?" In *Cultural Perspectives on the Mathematics Classroom*, edited by Stephen Lerman, pp. 85–97. Dordrecht, Netherlands: Kluwer Academic Publishers, 1994.

Pinxten, Rik, Ingrid van Dooren, and Frank Harvey. *The Anthropology of Space*. Philadelphia: University of Pennsylvania Press, 1983.

Vandenberg, Stephen G., and A. Ralph Hakstian. "Cultural Influences on Cognition: An Analysis of Vernon's Data." *International Journal of Psychology* 13, no. 4 (1978): 251–79.

von Glasersfeld, Ernst. *Radical Constructivism: A Way of Knowing and Learning*. London: Falmer Press, 1995.

Hands-On, Student-Structured Learning for American Indian Students

7

Claudette Bradley

Among Indian children play and learn amidst the daily activities of adults, who are engaged in subsistence-level work (Bradley 1984; Butler 1974; Guilmet 1984). Hours are spent observing adults before a child indicates readiness for participating and learning the same activities; these activities require skills that are important to learn for adult life.

Tlingit (pronounced Klin´-ket) people, for example, are historically located in southeast Alaska. Their children observe and participate in subsistence-level hunting and fishing with the adults, who manipulate the environment with great skill, requiring understandings of mathematical concepts and physics. For example, Tlingit adults construct wooden boats and fish-drying sheds, weave fishnets that must withstand the velocity and force of water, select rifles to suit the size and speed of the waterfowl and mammals they hunt, and sew clothing that will maintain their body temperature in rapidly changing environments (Guilmet 1984).

In Navajo country, children observe adults weave rugs, weigh corn, classify rocks, make silver jewelry, trade and buy goods, and care for sheep (Butler 1974). Rugs have two-dimensional surfaces on which the weaver creates a representational design with many geometric patterns. Likewise, jewelry patterns are geometric, with many symmetries.

RESEARCH ON CLASSROOM PRACTICE

Research (Penick 1981) in classrooms supports the effectiveness of hands-on, student-structured learning. John E. Penick, of Florida State University, studied the impact of teacher-structured classes versus student-structured classes that used the same lesson plans and manipulative materials. In the teacher-structured (TS) classes, the students manipulated objects under the direction of the teacher, who continually praised or rejected the students' behavior. In the student-structured (SS) classes, the students manipulated objects with no specific directions or evaluative feedback (Penick 1981).

The students in SS classes consistently exhibited on-task lesson-related behaviors more often than the students did in TS classes. The students in SS classes spent more than 90 percent of their time on lesson-related activities; disruptive behaviors decreased significantly; and productive, on-task, lesson-related behaviors increased. The students in TS classes spent 75 percent of their time on lesson-related tasks. Both sets of classes scored equally well on tests of verbal creativity, but the SS students scored significantly higher in figural creativity. The greatest impact of hands-on learning in SS classes occurred with the lower-ability students. The performance of these students increased more than that of any other ability group in either the TS or SS classes, on the basis of scores on the pretest and posttest (Penick 1981).

Penick's study suggests that all students may learn better by manipulating objects in a student-structured environment than in a teacher-structured one. If so, the effective teaching of American Indian students should replicate the student-structured approach in mathematics classrooms as a vehicle for developing formal mathematical thinking.

AISES MATHEMATICS CAMP PATTERNS CLASS

To test Penick's hypothesis, a Patterns class was designed for twenty-six twelfth-grade students who attended the American Indian Science and Engineering Society (AISES) Mathematics Camp held at Oklahoma State University (OSU) in June 1994. The students were from twelve tribes in the United States (Navajo, Mohawk, Pueblo, Comanche-Kiowa, Lumbee, and Sioux, to name a few), including two from Alaska (Tlingit and Blackfoot). All the students were high achievers in mathematics, earning grades of A or B in their home school; thirteen had attended AISES camps during the previous five years; all were college bound in fall 1995.

During the camp session the students were divided into one section of twelve and one of fourteen. Each section attended three mathematics classes in the morning. All the students convened in the computer room for a series of computer workshops in the afternoon. Of the three mathematics classes, the Patterns class was the only student-structured one. Each week the students were reorganized into eight small groups of three or four students. The small groups received three to four mathematics problems, each written on five-inch-by-seven-inch cards (Kohl 1987; Musser and Burger 1994). Each set of cards contained at least one problem requiring the students to make a connection with American Indian culture (generally baskets or rugs); a few sets contained problems that required connections with nature or the outdoors. Each group selected one problem from the set of cards to solve using hands-on materials. Each group submitted a plan that included—

1. a written statement of the problem;
2. a list of the materials needed;
3. a plan for the solution;
4. a list of the specific tasks given to each group member.

The materials were purchased and gathered by the mathematics department staff. The students were given ample time to solve the problems using hands-on materials. They created graphs, tables, diagrams, and descriptions for a poster display and organized themselves to make an oral presentation to the class. The posters defined the problem, described the procedure or solution, and displayed the results. Every poster had a visual portion: photographs, drawings, diagrams, charts, tables, or graphs. The poster was hung on the wall for easy reference during the presentation. The posters remained on the wall until the end of the camp.

Week 1

The eight small groups worked on five different subsets of the following seven problems during the first week: fractals, symmetry, transformations, pentominoes, binomial expansion, tessellations, and symmetry in tessellations.

Fractals

In the Fractals problem (fig.7.1a), students constructed fractal patterns using centimeter cubes (see fig. 7.1b). The ideas for the patterns came from several sources (Berger 1989; Devaney 1990; Kohl 1987). The problem required the students to—

1. define fractals;
2. illustrate three fractal patterns on paper;
3. construct three-dimensional fractal patterns with centimeter cubes and photograph the constructions;
4. write a description of the three-dimensional fractal constructions;
5. identify fractal patterns in American Indian designs.

The students were given literature on fractals from Kohl (1987, pp. 72–75), Devaney (1990), and Berger (1989). They received instruction on fractals and on using the computer to make fractal patterns in one of their afternoon computer workshops. This problem was a nice complement to that experience.

After reviewing the literature, the students selected three patterns. The first was an L shape made with three one-centimeter cubes. The L became the unit for the next level, and so on, until the L shape had generated a pattern similar to the Sierpinski triangle, which we could call a "Sierpinski pattern" (see fig. 7.1c).

(b)

Fractals

1. Define *fractal*.
2. Illustrate three fractal patterns on paper.
3. Construct three-dimensional (3-D) fractal patterns with centimeter cubes. Photograph your constructions.
4. Write a description of your fractal 3-D constructions.
5. Find and identify fractal patterns in American Indian designs.

(a)

(c)

Fig. 7.1. The Fractals problem and students' "Sierpinski pattern" made with centimeter cubes

The second pattern was generated by arranging four one-centimeter cubes in a T shape. The T shape became the unit for the next level, and so on, until the T shape generated the Sierpinski pattern in figure 7.1d.

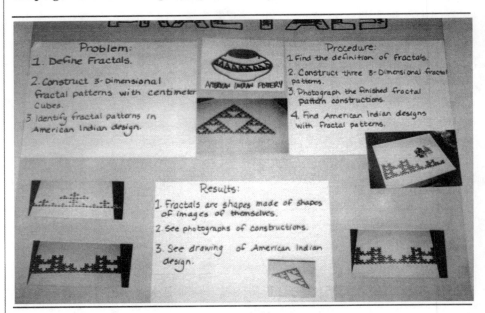

Fig. 7.1d. "Sierpinski pattern" generated by a T shape

The third pattern was generated by arranging five centimeter cubes in a U shape. The U shape became the unit for the next level, and so on, until the U shape generated the Sierpinski pattern in figure 7.1e.

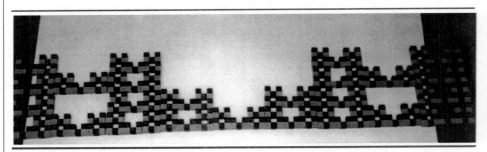

Fig. 7.1e. "Sierpinski pattern" generated by a U shape

The students photographed their work for their poster (fig. 7.1f), and, using their prior knowledge of pottery, they drew an American Indian pottery jug with a level-2 Sierpinski pattern. A level-3 Sierpinski pattern exists in a Maidu tribal basket (Appleton 1971). Harmsen (1990) and Turnbaugh and Turnbaugh (1986) describe students' explorations of rug and basket designs.

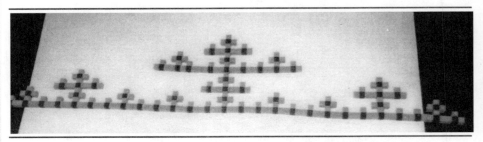

Fig. 7.1f. Photograph of a three-dimensional fractal pattern for students' poster

Symmetry

The Symmetry problem (fig. 7.2a) required the students to define the four basic *isometries*, or motions that preserve length: reflection, rotation, translation, and glide reflection. They also needed to be able to illustrate combinations of the two basic planar symmetries—point and line—and select examples of American Indian designs that illustrate combinations of these symmetries. The students used Geo Pieces, twenty-two different plastic polygons and circles in four colors, to construct symmetric figures (fig. 7.2b). The Geo Pieces were traced onto construction paper for cutouts, which were pasted on the display board. (fig. 7.2c) During their presentation, the students pointed to a poster of a Navajo rug hanging on the wall when discussing symmetry in American Indian designs.

(b)

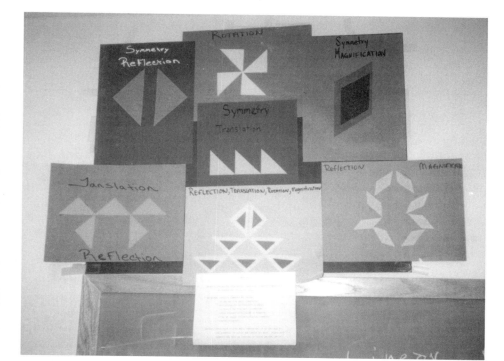

(c)

Symmetry

Can we identify isometries in designs?

1. Define the four basic isometries. Use Tangrams or pattern blocks to illustrate the isometries.

   (a) Reflection (line symmetry)
   (b) Rotation
   (c) Translation
   (d) Glide reflection

2. Illustrate combinations of two isometries.

3. Select samples of American Indian designs, and identify the combination of isometries.

(a)

Fig. 7.2. Symmetric figures made from construction-paper cutouts

Transformations

This problem required the students to plot three points on graph paper and connect them to form a triangle. They then plotted the image of the points under six given transformations and connected the points to form triangles. For each example, the students were asked to identify the symmetries in each transformation displayed (see fig. 7.3). The students discovered that tracing paper was valuable in determining the symmetry. The original triangle formed by connecting the three original points was traced, then the tracing paper was rotated for a rotation, moved horizontally or vertically for a translation, flipped for a reflection, or moved horizontally or vertically and flipped for a glide reflection. This technique was very effective and valuable in solving other problems, also.

Transformations

For each transformation given below, plot △*ABC*, with *A*(2, 3), *B*(–1, 4), and *C*(–2, 1), and its image. Decide whether the transformation is a translation, rotation, reflection, glide reflection, or none of these. The notation *T*(*x*, *y*) denotes the image of the point (*x*, *y*) under transformation *T*.

(a) $T(x, y) = (y, x)$
(b) $T(x, y) = (y, -x)$
(c) $T(x, y) = (x + 2, y - 3)$
(d) $T(x, y) = (2x, -y)$
(e) $T(x, y) = (-x + 4, -y + 2)$
(f) $T(x, y) = (-x, y + 2)$

(Musser 1991)

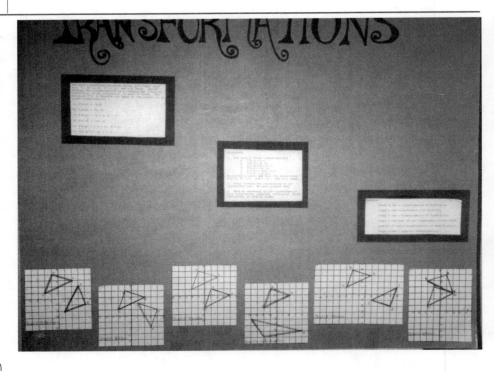

Fig. 7.3. The Transformations activity (Source: Musser [1994, p. 841])

Pentominoes

While the transformations group was discussing the triangles and their images, a student in the symmetry group asked whether they were doing the same problem. The response was no, but the connection was made. One group was looking for symmetries in designs, and the other was looking for symmetries in functions. A third group continued the connection of symmetry in pentominoes. That group was required to identify line and rotational symmetries among the twelve pentomino pieces.

Binomial Epansion

The students received literature about the binomial-expansion formula. (Cohen 1991). They used centimeter cubes to construct two-dimensional (2-D) and three-dimensional (3-D) examples of the binomial expansion of $(2 + 3)^2$ and $(2 + 3)^3$. The completed constructions and the separation of the constructions were photographed (see fig. 7.4). Then the students identified the separations with the numbers in Pascal's triangle. The students did not find the 2-D and 3-D characteristics sufficient to justify the connection of the binomial expansion with Pascal's triangle, so the instructor explained that (2 + 3) times (2 + 3) means five rows of five cubes composed of a 2×2 square of red cubes, a 3×3 square of blue cubes, a 2×3 green rectangle, and a 3×2 green rectangle. This arrangement gives a 2-square plus two 2×3 rectangles plus a 3-square.

Tessellations

The students were given literature from Kohl (1987) on tessellations and heard a lecture on Escher tessellation at the Museum of Science in Oklahoma City, which gave the students background information for the activity. The tessellations group used Geo Pieces to construct regular, semiregular, and Escher-type tessellations and drew tessellated patterns on dot paper. The dot paper was useful in enabling the students to draw a tessellated dog pattern. They also looked for tessellated patterns in American Indian designs and found a tessellation in one Navajo rug pattern (Harmsen 1990) (see fig. 7.5) and in several baskets (Turnbaugh and Turnbaugh 1986).

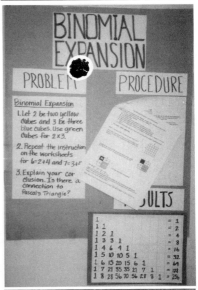

Fig. 7.4. The Binomial Expansion project

Fig. 7.5. Students' work on the Tessellations activity

Symmetry in Tessellations

This problem involved some research. The students were given four tessellated patterns (see fig. 7.6) and asked to find the lines of symmetry and the centers of rotation along with the angles of rotation. They used mirrors and tracing paper to find the lines of reflection and the centers of rotation. The students described the problem, procedure, and results on their display board and demonstrated the use of the mirrors and tracing paper in their oral presentation. This project was a nice tie-in with the Tessellation problem and the Symmetry problem.

For each tessellation, find rotations that map the tessellation onto itself. Indicate the centers of rotation and possible angles (less than 360°) of each rotation.

For each tessellation, find four lines of reflectional symmetry.

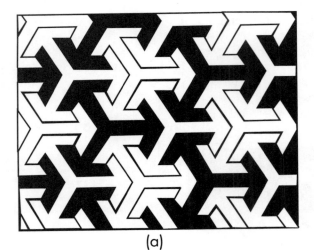

(a)

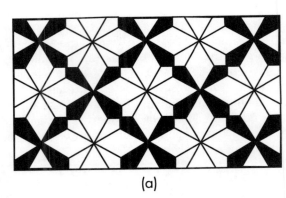

(a)

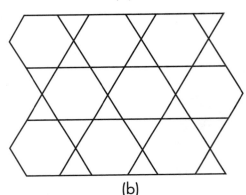

(b)

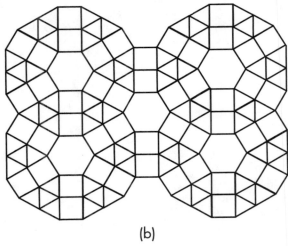

(b)

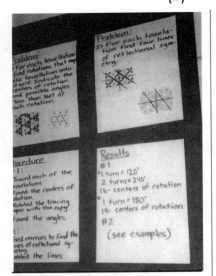

Fig. 7.6. The Symmetry in Tessellations activity (Source: Musser [1994, p. 838]).

Week 2

To enhance further the learning experience of the students listening to the oral presentations, an additional task was added the second week. Each group was asked to design a problem for the other groups to solve that was similar to the one they were presenting. This task was very successful.

Cross Sections of a Cube

The students made eight clay cubes and cut the cubes to make cross sections in the shape of a square, an equilateral triangle, a nonsquare rectangle, an isosceles triangle, a trapezoid, a regular hexagon, a pentagon, and a parallelogram. The students drew cubes on graph paper and colored the cross sections red for the display board (see fig. 7.7). The students worked many hours on this problem but were not able to find a similar problem for the other students to solve.

The clay was a delightful discovery and an effective manipulative. The students struggled to find the cross sections with pentagonal and hexagonal faces. The clay allowed them to experiment and make the solution real and meaningful.

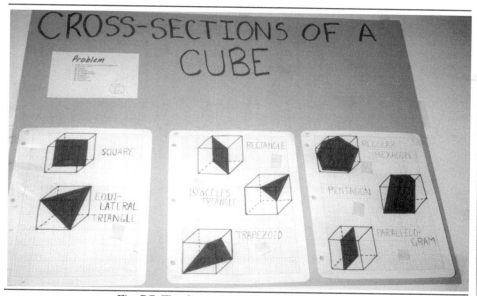

Fig. 7.7. The Cross Sections of a Cube display

Triangulated Structures

Straws and common pins were used to fashion a triangle, a quadrilateral, a quadrilateral with one diagonal, and a pentagon. Using gentle pressure with their hands, the students tried to collapse the structure to test for stability. The quadrilateral and pentagon collapsed immediately. The triangle and the quadrilateral with one diagonal held their shape and were considered stable.

The students constructed a truss (composed of 3-D tessellated tetrahedrons) with toothpicks and clay (see fig. 7.8). The truss was a very stable construction with many triangles in every direction. The students challenged their classmates to make a dome with toothpicks and clay. They compared the dome with the truss and found the truss to be more stable. They then discussed the result.

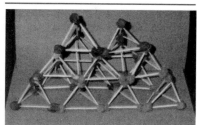

Fig. 7.8. The Triangulated Structures display

Numbers on a Cube

The students constructed a large cube with centimeter cubes. They wrote the numbers 1, 2, 3, 4, 5, 6, 8, 9, 10, 11, 12, and 13 on circle stickers. The stickers were placed on the edges of the cube and manipulated until the four edges on any face added to 28 and the three edges adjacent to any vertex added to 21. During its oral presentation, the group was not able to present a similar problem to challenge the other students.

Squares inside Squares

The students made copies of a large square region. Using the numbers 7, 8, 11, 12, 13, 14, 15, and 16, they divided one square each into those numbers of smaller squares. The students asked the class to solve a similar problem for the number 29.

Length versus Area (Map of Oregon)

The students plotted twelve points given on a card and connected the points consecutively. The resulting shape formed the outline of the state of Oregon. The students were asked first to double the distance between consecutive points and then to determine the coordinates of the corresponding twelve points if the area of the map is doubled. The students had trouble with this problem. When they doubled all the coordinates of the map, the resulting area was four times that of the original outline of Oregon. The students discovered the solution after several hours of discussion and trial and error (see fig. 7.9). The students asked their classmates to describe the solution if they wanted to simply triple the area.

Equilateral Triangles

The students used triangle dot paper to draw and count the number of equilateral triangles with whole-unit sides in a $6 \times 6 \times 6$ and an $8 \times 8 \times 8$ triangle. They had a $2 \times 2 \times 2$ plastic triangle from the Geo Pieces to help them count the number of $2 \times 2 \times 2$ triangles in each of the given triangles. For the other sizes, tracing paper worked well. Since tracing paper can be flipped over, the students could easily count the upside-down triangles. During its presentation, this group asked the class to solve the problem for a $5 \times 5 \times 5$ triangle.

Length versus Area (Map of Oregon)

Plot the following points and connect them to make a map of Oregon.

A(–7, 5)	E(3, 4)	I(6, –2)
B(–5, 5)	F(6, 4)	J(6, –7)
C(–4, 3)	G(7, 3)	K(–8, –7)
D(0, 3)	H(5, –1)	L(–8, –3)

Connect A to B to C, and so on, and connect L to A.

(a) Make a map of Oregon in which all lengths are doubled from your original map.

(b) Make a map of Oregon whose area is twice that of the original map. (This is not the same map as in part (a). Draw a picture, and use coordinates.

(Musser 1994)

Fig. 7.9. Length versus area (Map of Oregon) activity (Source: Musser [1994, p. 803])

Map Coloring (Four-Color Problem)

The students reviewed the literature on the four-color problem in Kohl (1987). They drew examples of the two-color problem, the three-color problem (using the six states that border Wyoming), and the four-color problem. They looked in Harmsen (1990) and Turnbaugh and Turnbaugh (1986) for examples of American Indian basket and rug designs that illustrate these problems. They drew patterns for other students to color and asked them how the concepts in the four-color problem apply to the patterns (see fig. 7.10).

Map Coloring

(Four-Color Problem)

1. Explain map coloring
2. Discuss and illustrate the two-color theorem
3. Discuss and illustrate three-color mappings.
4. Discuss and illustrate the four-color theorem
5. Discuss the limitations of Indian designs in each of the following cases.
   (a) Using only two colors
   (b) Using only three colors
   (c) Using only four colors

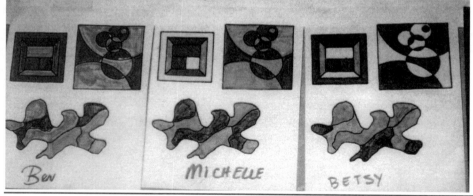

Fig. 7.10. Patterns involving the Four-Color Problem activity

Height of a Tree

The students determined the height of a tree in the park by two methods: using a student's shadow and using a mirror (see fig. 7.11). With string and masking tape, they were able to measure lengths, and they then solved a proportion to determine the height of the selected tree. They compared the length of a student's shadow to the length of the tree's shadow or created similar triangles by standing a measurable distance from a mirror on the ground until the top of the tree could be seen. These students were not able to think of an alternative problem to give the other students during their presentation.

Evaluation of the Patterns Class

The students took the eighteen-item Mathematics Attitude Questionnaire (Brown 1977) both before and after the Patterns class. Nine items expressed positive statements about mathematics, and nine expressed negative statements. For each item, each student circled one of the following responses:

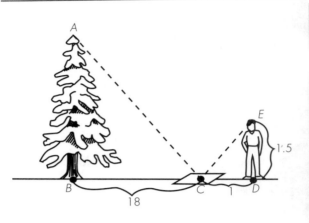

Height of a Tree

At a particular time, a tree casts a shadow 29 m long on horizontal ground. At the same time, a vertical pole 3 m high casts a shadow 4 m long. Calculate the height of the tree to the nearest meter.

Another method of determining the height of an object uses a mirror placed on level ground. The person stands at an appropriate distance from the mirror so that he or she sees the top of the object when looking in the mirror.

(a) Name two similar triangles in the diagram. Why are they similar?

(b) Find the height of the tree if a person 1.5 meters tall sees the top of the tree when the mirror is 18 meters from the base of the tree and the person is standing 1 meter from the mirror.

(c) Determine the height of a tree or building outdoors.

Fig. 7.11. Students' work on the Height of a Tree project (Source for parts (a) and (b): Musser [1994, p. 717])

SA—Strongly Agree, A—Agree, N—Neither Agree nor Disagree, D—Disagree, SD—Strongly Disagree. A comparison of the scores on the pretest and posttest (table 7.1) reveals that only one student changed from disliking mathematics to the neutral category. On her evaluation form, she wrote, "The class was neat."

Table 7.1
Mathematics Attitude Questionnaire Results

Degree of Feeling toward Mathematics	Number of Students	
	Before Camp	After Camp
Strongly Like	4	4
Like	15	15
Neither Like nor Dislike	1	2
Dislike	1	0
Strongly Dislike	0	0

In their evaluation of the class, the students listed the three projects that they had worked on over the three-week period. The two projects liked most by three members of a group were Tessellations and Transformations. The four projects liked the most by at least two members of a group were Height of a Tree, Symmetry, Fibonacci in Nature, Pentominoes, and Symmetry in Tessellations. The students listed three other projects that interested them but that they had not worked on. Table 7.2 lists the six most interesting other projects. All six of these projects were solved in the second week. The absence of projects from the third week is understandable, since only one was completed and none were presented to the class. The absence of additional projects from the first week is curious. For the first four listed, the students had made up challenge problems for the other students to solve during the oral presentation. Perhaps this addition to the problem enabled the students to remember the problem more easily and appreciate it more.

Table 7.2
Most Interesting Projects Selected by Students

Project	Number of Students Selecting Project
Triangulated Structures	8
Map Coloring/Four-Color Problem	8
Squares inside Squares	6
Equilateral Triangles	5
Numbers on a Cube	4
Cross Sections of a Cube	2

On a twelve-item section of the evaluation form, the students circled the response that most closely represented their feelings. The responses with their numerical codes are shown in table 7.3. For the most part, the students liked the activities of the hands-on, student-structured mathematics class. They gave the highest ratings to the hands-on materials, working in groups, and participating in a student-centered mathematics class. The two lowest ratings—given to "oral presentations" and "collaborating on the problem"—fell in the neither-liked-nor-disliked range.

Table 7.3
Students' Ratings of the Twelve Components of the Patterns Class

SL—Strongly Liked (2)
 L—Liked (1)
 N—Neither Liked nor Disliked (0)
 D—Disliked (–1)
SD—Strongly Disliked (–2)

Item	Average Response (Numerical Code)	Interpretation
Using hands-on materials	1.35	Liked
Working in groups	1.30	Liked
Student-structured class	1.27	Liked
Making a poster	1.18	Liked
Creating an experience	1.09	Liked
Observing posters of others	1.04	Liked
Selecting problems	0.96	Liked
Watching oral presentations	0.91	Liked
All working on problems	0.70	Liked
Collaborating on the solution	0.65	Liked
Collaborating on the problem	0.48	Neither
Oral presentations	0.43	Neither

Of the twenty-three students who responded to the evaluation, twenty wrote comments on the question "What aspects of the Patterns class did you like and why?" Four students said that they enjoyed working in groups, three enjoyed selecting the problems and working with patterns, four said that the projects were fun and interesting, and five appreciated the hands-on materials. Only two students offered suggestions to improve the class. One wanted "more time to spend and know more about it (the problem)" but also offered, "I like it very well, otherwise." The other wanted "more detail[ed] explanation of work to be done."

In conclusion, the literature strongly supports the hypothesis that hands-on, student-structured environments are highly effective for American Indian students. Certainly, the students in this study demonstrated an appreciation for student-structured environments with hands-on learning experiences. In such a mathematics class, students can make connections among mathematical topics as well as to their culture and environment, as recommended in the NCTM's *Principles and Standards for School Mathematics* (2000). The hands-on activities make cognitive bridges between the knowledge base of Indian students and formal mathematical thinking.

REFERENCES

Appleton, Le Roy H. *American Indian Design and Decoration.* New York: Dover Publications, 1971.

Berger, Marc A. "Images Generated by Orbits of 2-D Markov Chains." *Chance: New Directions for Statistics and Computing* (spring 1989).

Bradley, Claudette. "Issues in Mathematics Education for Native Americans and Directions for Research." *Journal for Research in Mathematics Education* 15 (March 1984): 96–106.

Brown, Ric. *Construct Validation of Attitudes toward Mathematics.* Research Report. Washington, D.C.: U.S. Department of Health, Education, and Welfare, National Institute of Education, 1977.

Butler, Katie B. "Walking in Navajo Footprints." Houston, Tex.: Houston University, Texas College of Education, 1974.

Cohen, Don. *Calculus by and for Young People: Worksheets.* Champaign, Ill.: Don Cohen, the Mathman, 1991.

Devaney, Robert L. *Chaos, Fractals, and Dynamics.* Menlo Park, Calif.: Addison-Wesley Publishing Co., 1990.

Guilmet, George M. "American Indian and Alaskan Native Education for High Technology: A Research Strategy for Creating Culturally Based Physical Science and Mathematics Education." Paper presented at the annual meeting of the Society for Applied Anthropology, University of Puget Sound, Tacoma, Washington, March 1984.

Harmsen, William D., ed. *Patterns and Sources of Navajo Weaving.* Denver: Harmsen Publishing Co., 1990.

Kohl, Herbert. *Mathematical Puzzlements.* New York: Schocken Books, 1987.

Musser, Gary L., and William F. Burger. *Mathematics for Elementary Teachers.* 2nd ed. New York: Macmillan Publishing Co., 1994.

National Council of Teachers of Mathematics (NCTM). *Principles and Standards for School Mathematics.* Reston, Va.: NCTM, 2000.

Penick, John E. "Teacher Behavior Does Make a Difference in Hands-On Science Classrooms." *School Science and Mathematics* 81 (November-December 1981): 46–47.

Turnbaugh, Sarah, and William Turnbaugh. *Indian Baskets.* West Chester, Pa.: Schiffer Publishing, 1986.

Culturally Relevant Assessment for the American Indian Student

Developing Mathematics Portfolios

8

Libby Quattromani

Joanna Austin-Manygoats

"The task challenging Native communities is to retain their distinct cultural identities while preparing members for successful participation in a world of rapidly changing technology and diverse cultures," according to the Indian Nations At-Risk Task Force (Gilliland 1995, p. 1). American Indian educators have expressed the essence of the American Indian educational experience, that is, surviving in a parallel universe. How do we move beyond the dark world of pedagogical oppression (Freire 1994) and into the light of culturally relevant curriculum, pedagogy, and assessment? What must we come to know and believe about best educational practices for American Indian children? If our instructional practices move beyond the traditional, how will we maintain congruence between pedagogy and assessment? This article describes using mathematics portfolios as a possibility for assessing American Indian students' achievement through cultural relevance and authenticity.

American Indian students more eagerly engage in mathematics learning when the concepts are presented through a culturally relevant curriculum (Gilliland 1995). The notion of a culturally relevant curriculum is not to suggest that somehow American Indian students should be educationally stereotyped as possessing one learning style or another. Case studies and current research hypotheses (Cleary and Peacock 1998; Swisher and Deyhle 1992; Gilliland 1995) support the notions that American Indian children have as many learning-style preferences as non–American Indians do but that difficulties occur because the ways in which American Indian children learn at home conflict with the ways in which schools teach (Cleary and Peacock, 1998, p. 157):

> How might recognition of this play out in practice? Classrooms need to integrate culture into curriculum to blur the boundaries between home and school. Schools need to become a part of, rather than apart from, the communities in which they serve.

The desire expressed in these observations, to support and respect our American Indian students and their families, has been the impetus for educational transformation. Consider the current state of assessment of students' mathematics achievement in your district, school, or classroom. What are your concerns? Do many teachers think that standardized tests have serious limitations, particularly with respect to their students' cultural and linguistic background? Is consistency with culture, national standards, curriculum design, pedagogy, and assessment evident; do conflicts result from curricula that

CULTURALLY RELEVANT MATHEMATICS CURRICULUM AND ASSESSMENT

continue to rely on culturally limited assessment practices; or both? We suggest that portfolio assessment constitutes a more comprehensive, culturally relevant process for mathematics assessment and, in many ways, addresses and avoids some of these conflicts. Therefore, if mathematics-portfolio assessment provides such an opportunity for curricular consistency, what does the process involve and what characterizes it? In other words, what is a mathematics portfolio?

WHAT IS A PORTFOLIO?

A portfolio is (*a*) an edited collection of artifacts, or materials, that provide a framework for the student to demonstrate knowledge, understanding, experiences, and learning processes; (*b*) not only an active record of a student's experiences but also an indicator of how goals have been achieved; (*c*) a tool to encourage students' self-analysis of academic and personal decisions; (*d*) evidence of verifiable learning outcomes; and (*e*) often a messy business.

Rationale

Teachers and learners are responsible for making myriad daily educational decisions. These decisions are the framework for developing students' mathematics portfolios. The driving force behind developing a working, authentic-assessment mathematics portfolio is the realization that education is a complex, multidimensional human experience that relies heavily on interactive processes. Historically, this experience occurs over time and is often recorded only in memory and relegated to one or two evaluative tests or worksheets with a primary focus on computational skills. In contrast, the mathematics portfolio is a vehicle for organizing students' artifacts that aid in the assessment of performance, creativity, participation, group interactions, decision-making skills, product development, and other essential learning experiences that are otherwise difficult to consider in any other holistic manner.

Because education is such a complex human experience, the nature of assessment must reflect the elements of such a complex process. The development of a mathematics portfolio provides educational stakeholders—students, parents, teachers, administrators, and others—with a tool for reflecting on this process. If we recognize the individual growth processes over time within a cultural context, then we will be better able to use a holistic student-assessment process while maintaining an overarching personal identity for American Indian students. NCTM's *Curriculum and Evaluation Standards for School Mathematics* (1989) provides the necessary guidelines for comprehensive curriculum development. Furthermore, the mathematics portfolio is a product—or more precisely, a collection of artifacts—that directly reflects the nature of the learning process and can lead to the improvement of both the student's achievement and the teacher's professional competence.

Purpose

The mathematics portfolio is designed to accomplish two purposes: (1) to provide samples of student mathematics achievement in problem solving and mathematical communication and (2) to provide indicators of the quality of the math program that the child is experiencing. (Abruscato 1993, p. 475)

It is very important that we focus not only on students' achievement but also on the quality of the mathematical experiences offered to the student. This precept is what we mean by the term *contextualized measures*—assessing students on the basis of actual mathematics program experiences, not those that are presumed by external measures.

Organizing Principles

As they embark on the process of portfolio assessment, teachers need to keep in mind several organizing principles:

- Learning occurs in diverse settings.
- Students have different styles of learning.
- Collaboration enhances educational growth in students.
- Self-reflection is necessary to improve teaching and learning.
- The evaluation of students should lead to changes in the curriculum for students' improved achievement.

Essential Questions for Criteria Development

As Abruscato (1993) describes, possible criteria for assessing the portfolio can be selected from a wide array:

Problem Solving

- How well does the student understand the problem?
- How does the student solve the problem?
- Why does the student solve the problem in a particular way?
- What observations, connections, and generalizations does the student make about the problem?

Communication

- What terminology, notation, and symbols does the student use to communicate mathematical thinking?
- What representations (graphs, charts, tables, diagrams, pictures, or manipulatives) does the student use?
- How clear is the student's communication of mathematical thinking and problem solving?

Quality of the Mathematics Program

- What is the student's attitude toward mathematics?
- What are the student's responses in a variety of mathematics content areas: number sense, number relationships, operations, estimation, measurement, and so forth?
- What sense of the student's empowerment do we perceive from evidence of curiosity, flexibility, risk taking, perseverance, and reflection?
- What instructional opportunities (e.g., group work, interdisciplinary work, the use of manipulatives, real-world applications, technology) are available to the student?

Conceptual Development

A mathematics portfolio's meaningfulness is enhanced and magnified by the degree to which it is based on and reflects the cultural community's educational beliefs.

It is important for teachers to become familiar with the cultural and linguistic patterns of American Indian students. Teachers must realize that there are many differences between American Indian tribes, and a wide range of differences between American Indian students and the white majority. The greatest stumbling block to full interpersonal acceptance is ignorance of a way of life which has ideals, habits, history, and even language different from what seems standard. It is clear that the successful completion of school by American Indian students may well depend on the teacher's ability to understand the culture.

—Jeanne Bearcrane, Crow
Teaching the American Indian

Without philosophical grounding and evidence of beliefs that are congruent with the American Indian cultural community, the portfolio tends to become a scrapbook. The reflective dimension of the portfolio makes it a powerful learning tool. Therefore, the power of the portfolio will be enhanced when its meaningfulness is congruent with the cultural community's educational goals and philosophy. After this congruence is established, the portfolio as a product begins to take shape.

Physical Design, or Format

Two parts of the physical development should be considered: (1) the way materials are organized and (2) the presentation or packaging of the entire portfolio. The physical format of students' portfolios can, and should, complement the conceptual base. The organization and presentation of the goals and experiences are crucial because they offer one other way for the students to explain themselves.

I think of two landscapes … one outside the self, the other within. The external landscape is the one we see.… The second landscape I think of is an interior one, a kind of projection within a person of a part of the exterior landscape.

—Barry Lopez
Crossing Open Ground

The Organization of Material

Portfolio materials can be organized according to the district, state, or NCTM Standards, as well as according to student goals. Despite this structural framework, considerable room is allowed for personal identity within the structure. Additionally, the design should include evidence of growth over time. The presentation, or packaging, of the entire portfolio requires considerable collaboration, creativity, and thought:

"All look your best," Peter warned them. "First impressions are awfully important." He was glad no one asked him what first impressions are, they were all too busy looking their best.

—Peter Pan
Peter Pan

Selection and Analysis of Artifacts

"She's in that state of mind … that she wants to deny something—only she doesn't know what to deny!"

—White Queen
Alice's Adventures in Wonderland

- *Best evidence:* In collaboration with other educational stakeholders, determine which materials can provide the best evidence of the beliefs, skills, understandings, and experiences that the student wants or needs to present.
- *Artifacts:* Artifacts are the materials selected, for example, (*a*) five to seven "best pieces"—among them at least one puzzle, one investigation, one application, and no more than two pieces of group work—listed in a table of

contents and including the solution and the work involved; (*b*) a letter to the portfolio evaluator; and (*c*) a collection of other pieces of mathematics work (Abruscato 1993, p. 476).

BEST
EDUCATIONAL
PRACTICES

Taylor (1996) states that teachers can create opportunities for students to make meaningful connections among mathematics, social studies, art, language arts, science, and physical education. These connections are at the core of integrated instruction, that is, instruction based on the belief that knowing is multidimensional and requires information from a variety of sources. Because content and procedures transcend individual disciplines, connected teaching is a natural way to teach and learn (Taylor and Quattromani 2001). Therefore, if we are truly to believe in a multidimensional learning experience, we can no longer rely exclusively on decontextualized assessment, that is, assessment by such external measures as standardized tests. Nor can we continue to rely on paper-and-pencil testing, whether contextualized (occurring within the context of students' day-to-day learning experiences) or decontextualized, as the *only* means for assessing students' achievement.

Consider the comprehensive work of Zemelman, Daniels, and Hyde (1993), which describes the thirteen characteristics of best practice: (1) child centered, (2) experiential, (3) reflective, (4) authentic, (5) holistic, (6) social, (7) collaborative, (8) democratic, (9) cognitive, (10) developmental, (11) constructivist, (12) psycholinguistic, and (13) challenging (pp. 7–8). They give credit to the National Council of Teachers of Mathematics for the tremendous contribution of its *Standards* documents published since 1988. "These frameworks and guidelines have shown other fields how learning goals for children can be described in Best Practice terms—progressive, developmentally appropriate, research-based, and eminently teachable" (p. 6).

Over the past twenty-seven years, the first author of this chapter has used portfolios as both formative and summative assessment tools. As early as September 1970, when she walked to her car as a first-year teacher and saw the "paper trail" of graded worksheets blanketing the neighborhood like an early snowfall, she realized that a better way had to be possible. Since that time, what was a very primitive portfolio has grown in scope and, she hopes, sophistication.

Mathematics Literacy and Problem-Based Learning

All children from the most to the least socially advantaged family circumstances, come to school knowing and doing various kinds of literacy. It follows that the kinds of literacy work and play that teachers and students begin to do in classrooms necessarily should look and sound different, given differing combinations of racial, ethnic, linguistic, and class backgrounds of actual sets of teachers and students. (Hiebert 1991, p. x)

Given the importance of problem solving, word problems, and other linguistic aspects of the mathematics curriculum, one must ask why students who do not have basic English-language skills are expected to understand complex concepts in the English mathematics register without specific, culturally based instruction. Additionally, English-language development is increased when it occurs specifically in the context of the mathematical terms, skills, and concepts. The need to feel competent before engaging in an activity was found in several studies of American Indian students.

Appleton (1983), Brewer (1977), Longstreet (1978), Wax, Wax, and Dumont (1964), and Werner and Begishe (1968) studies of Navajo, Sioux, and Yaquis children found that many of these students were ill at ease and preferred to watch an activity and would not physically engage in it until they were confident that they could perform it. Non-Indian children, however, often try again until they succeed. (Cleary and Peacock 1998, p. 157)

Therefore, American Indian students need to see a relationship between a discipline (e.g., mathematics) and how the discipline will apply (e.g., problem posing, problem solving) to their own culture before developing the facility to generalize this knowledge to a variety of circumstances. This finding is directly related to the Navajo culture through Life-Way instruction, or in Diné, *iiná*. Furthermore, to forge literacy connections for American Indian students, we, as educators, must recognize and comprehend content-specific mathematics language to impart full "mathematics literacy" for problem solving and related skills.

For the American Indian student, opportunities for experiencing authentic curriculum and assessment through problem-based learning (PBL) provide cultural relevance during the crucial early learning phases. PBL is an apprenticeship in real-world problem solving (Abruscato 1997, p. 41). It starts with encountering an actual problem or simulation. From the outset, students are assigned or choose a role or stake in the problem. Their role gives them power, authority, responsibility, and an identity during the unit of study. The following PBL scenarios are examples designed for Navajo students:

Example 1

In the Navajo culture, estimating the amount of grazing land needed to support a given number of sheep and calculating the number of people that can be fed from this flock for a Navajo wedding ceremony would be a culturally relevant learning experience. Drawings, graphs, data sheets, algorithms, and sociocultural interviews, as well as other processes and products, could serve as documents or artifacts in the students' mathematics portfolios.

Example 2

In the Navajo tribal culture, as in many American Indian tribal cultures, an excellent source for cross-cultural, interdisciplinary experiences is the calendar. The integration among the disciplines of science, mathematics, language arts, health, history, and art is rich with learning connections for students. Consider the comparisons, shown in figures 8.1, 8.2, and 8.3, between the Navajo and English calendars told to Becenti by the Navajo elders with Navajo written language by Manygoats.

Herein lies an opportunity to chart weather patterns from the past and present, checking for accuracy. This PBL experience allows the American Indian student multiple opportunities for implementing the same problem-solving strategies that we use in the adult world, including studying the life cycle of plants and animals in relation to human beings, making observations, creating hypotheses, conducting investigations, collecting information, defining issues, evaluating one's own thinking, developing possible solutions, and selecting solutions that fit our construction of the problem. During PBL, students explore various domains of knowledge to find information about questions and issues that perplex them. They make progress, and they run into blind alleys. They might have to rethink their strategies and collaborate with their classmates as they modify their hypotheses, questions, and searches for new information. Throughout the process, students are building new knowledge bases, refining their inquiry skills, and providing assessment information for their mathematics portfolios.

Navajo	English	Descriptions Associated with Months
Ghąąji	October	Added to summer months; beginning of the Navajo year. This is the month of the coyote. Winter is approaching; seasonal change from hot to cold.
Nítch' its'ósí	November	Wind—the wind is crisp, somewhat penetrating; wear warm clothing.
Nítch' itsoh	December	Harsh, cold wind. Absence of snakes, bears, spiders, and thunderstorm or lightning activity. Winter games and storytelling: creation stories and sacred legends.
Yas Niłt'ees	January	Top layer of snow melted and frozen; it looks like it has been fried. It is storytelling time. No thunderstorms.

Fig. 8.1. The weather months

Navajo	English	Descriptions Associated with Months
'Atsá Biyáázh	February	Eaglets hatch during this month.
Wóózhch'įid	March	First cry of eaglets. The sound of the wind.
T'ą́ąchil	April	Growth of eaglets' wings. First growth of plant sprouts in spring (Daan).
T'ą́ątsoh	May	Larger feathers of eaglets grow. First sign of perennials. Larger leaves mature.

Fig. 8.2. The months of the stages of the eagles

Navajo	English	Descriptions Associated with Months
Ya' iishją́ą́shchilí	June	Some ripening of early crops. Sowing early crop seed.
Ya' iishystsoh	July	Wild plant food is ripe. Some also used in ceremonies. Sowing late crop seeds.
Bini' ant'ą́ąts'ósí	August	Early harvest; corn is ready for eating.
Bini' ant'ą́ątsoh	September	Late harvest; all planted crops ripen. Recognition of boyhood and girlhood.
Bini' na'a;' ą́ąshįį	In between	In-between months, necessary to account for thirteen moons of the year.

Fig. 8.3. The months of the harvest

Learning Styles

Students taught through their preferred learning styles (*a*) achieve more academically than their counterparts, (*b*) are more interested in the subject studied, (*c*) like the way the subject is taught, and (*d*) want to learn other school subjects in the same way (Lazear 1994). Each particular way of knowing and problem solving is found in every culture, regardless of socioeconomic and educational conditions. Although certain skills are more highly developed in some cultures than in others, and in some distinct groups within a given culture, the occurrence of an intelligence is nevertheless universal. Through inherited abilities and through methods of instruction, children realize their preferred learning styles early in their development. Learning styles are varied, just as differences are evident among the traditional cultures of tribal groups (Gilliland 1995). However, nothing that is said applies to all tribal groups or to all individuals within any one tribal group. A beneficial approach, therefore, matches learning styles with varied teaching styles. Furthermore, people exhibit many forms of intelligence—many ways by which they know, understand, and learn about the world. Most of these forms go beyond those that dominate Western culture and education.

> Extensive research has shown that educators have differing expectations of students depending on the students' race, ethnicity, and gender. These studies have provided consistent data demonstrating stark disparities of class, caste, and entitlement in educators' interactions with students. Interactions based on poor expectations clearly lead to devastating consequences for students, in terms of both academic performance and self-image. (Lindsey, Robbins, and Terrell 1999, p. 116)

As educators, we must examine our instructional strategies so that we reach all students, not just those who are language oriented. This statement does not mean that we must accommodate all learning styles simultaneously but rather that we should vary our day-to-day practice so that at least two or three styles are acknowledged in a given lesson (Quattromani and Taylor 2001). An area of potential conflict in teaching American Indian children is the clash between the learning styles to which they have been exposed at home and those used in the classroom (Gilliland 1995). A large percent of American Indian students learn most easily by observing, imaging, and reflection (Gilliland 1995). Because teachers should build skills through active participation within cultural-environmental constructs, they should learn as much as possible about the instructional methods used in the homes of their students.

> As culturally proficient educators, we can strive to overcome these obstacles to learning through programs that provide models by which educators can learn of verbal and nonverbal behaviors that project the cultural expectation that all students can learn, thus providing them with equal opportunity in the classroom. (Lindsey, Robbins, and Terrell 1999, pp. 116–17).

In the traditional way of the American Indian, children are taught from understandings of the whole picture (concepts and ideas) to the details (skills); whereas in the non-Indian culture, we have historically taught from the details to the whole. Not until the late 1980s and the 1990s did the broader educational arena begin to teach from a concept-based model of whole-to-part.

Additionally, educators should recognize that in American Indian communities, children learn from one another as siblings and older companions. American Indian students tend to work together effectively and are more cooperative than non-Indian students. Some evidence indicates that classrooms having a cooperative rather than a competitive learning environment meet the unique learning needs of many American Indian students. "Cohen (1986), Slavin (1983), and Stahl and VanSickle (1992) found that many minority students learn best when cooperative teaching is used" (Cleary and Peacock 1998). "Cooperative learning organization provides a setting for cooperative study than can be employed in combination with many approaches to teaching" (Joyce, Weil, and Calhoun 2000, p. 451).

The Relationships among Cultural Influences

Figure 8.4 is a graphic representation of the interrelatedness of the elements in the American Indian environment that influence the educational process. This model should be considered in designing mathematics-portfolio assessments. All elements should be included, and their relationships should be reflected in the portfolio.

The primary educational elements in the learning environment include (*a*) the student; (*b*) the teacher; (*c*) the family, guardians, dormitory counselors, and the like; and (*d*) peers. In Navajo, educational influences are collectively called *Ké.* Communication initially occurs among educational stakeholders in this level of interaction. The secondary educational elements include (*a*) the tribe,

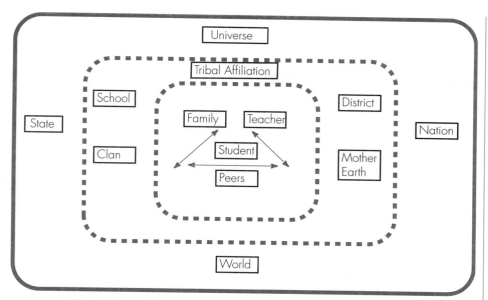

Fig. 8.4. A model of the influences in American Indian education

(*b*) the tribal clan, (*c*) the school, (*d*) the district, and (*e*) Mother Earth. This group of educational influences clearly reflects the duality of the American Indian student's education. The tertiary educational elements are further removed from the day-to-day life of the student and include (*a*) the state, (*b*) the nation, (*c*) the world, and (*d*) the universe.

The educational stakeholders provide not only expected standards for performance but also essential, more comprehensive evidence of students' achievement. In this collaborative manner, they become the sources of data, transforming assessment from a one-dimensional, teacher-centered process to a multidimensional, student-centered or culture-centered process.

Next we explore the planning in implementation of the mathematics portfolio.

MATHEMATICS-ASSESSMENT PARADIGM SHIFT

Think of the best mathematics student you ever had, then try to describe the characteristics of that student. More than likely, teachers would describe very different characteristics of their "best" mathematics student. When trying to identify the characteristics of our best students, we amaze one another with the variety of our descriptions. We use terms like *focused, divergent, creative, well-organized, demanding, caring, spontaneous,* and *methodical,* often in the same sentence. We are perplexed by the ambiguous and often contradictory nature of our descriptions.

Additionally, when we consider learning styles, cultural characteristics, gender, and linguistic development, the issue of assessment becomes even more complex. We are confounded when we try to use evaluation as a way to encourage the educational growth of our students. Ultimately, we agree that the "best" mathematics student presents a living educational paradox. If we agree that this descriptive paradox exists, then we are forced to consider alternative assessment methods that reflect the nature of this paradox. We must develop assessment strategies that are designed specifically to encompass as many of the attributes of our students as possible rather than just the isolated and competitive ones. Figure 8.5 compares traditional assessment practices with the more

Traditional Evaluation	Portfolio Assessment
Culturally limited	Culturally relevant
Decontextualized	Contextualized
Standardized	Individualized
Computational learning	Problem-based learning
Teacher centered	Student centered
Organized	Organized
One-dimensional	Multidimensional
Student isolation	Student collaboration
Quantified data and short answers	Learning styles
Data driven	Data driven
Content and product driven	Process and product driven
Isolated knowledge	Holistic knowledge
Risk of philosophical dissonance	Philosophically congruent
Reliable	Interrater-reliability concerns
(within narrow cultural parameters)	
One-shot, isolated event	Educational growth over time

Figure 8.5. A comparative analysis of traditional evaluation and portfolio assessment

authentic portfolio-assessment process. The portfolio-assessment model parallels traditional American Indian teaching and learning situations.

Figure 8.6 outlines the roles and responsibilities of the primary stakeholders in a student's education. The matrix presents possible considerations in the development of a culturally relevant mathematics-portfolio assessment program. Certainly, many more elements could be added or subtracted. The primary consideration is that the portfolio reflect the experiences of the learner within his or her learning environment.

Figure 8.7 shows an example of a student's portfolio plan, including her goals and self-evaluation. Figure 8.8 presents questions for the teacher to aid in the portfolio-planning process, and figure 8.9 lists questions to help the teacher plan outcome rubrics.

THE JOURNEY BEGINS

"The real challenge to the success of the portfolio system will be the support of teachers" (Abruscato 1993, p. 477). The process must be developed in a shared-decision-making model wherein teachers are supported with materials and training for maximum ownership. If teachers perceive that this approach is just another "mandate from above," the resistance-to-change phenomenon will be very strong and the energy for change will be drained. Casta (cited in Graves [1992]) reminds us, "We must constantly remind ourselves that the ultimate purpose of evaluation is to enable students to evaluate themselves."

Because grades seem counterproductive for many American-Indian students, performance assessment seems a much more productive way to go. Reservations that have their own schools may have some autonomy over this process. When a community decides what they think it is important for their students to know before they leave school, there can be lists of outcomes …, and assessment of whether students have met school expectations an be authentic, shown by the end results of students' endeavors. What you see or hear students do is what they can do. When such lists are publicly available, the mystery of grades and of lack of success no longer exists. What students can do and what they need to be able to do is abundantly clear. (Cleary and Peacock 1998, p. 243)

The opportunity for fundamental change in the way assessment of American Indian students is conducted has arrived. "Educators must consider the implications of curriculum and instruction within the context of the culture of students" (Stokes 1997, p. 576). In contract schools, public schools, and Bureau of Indian Affairs Schools, many educators have recently been engaged in developing curricula based on the values of their American Indian communities. If the educational community embraces the concept of culturally relevant curricula and assessment for American Indian students, then the possibility of students' improved achievement and a sense of pride among all educational stakeholders in the American Indian community does exist.

(References appear on p. 98.)

Considerations in Developing a Portfolio-Assessment Program

Roles and Strategies of the Educational Stakeholders

Considerations in Developing a Portfolio-Assessment Program	The Teacher— (As professional educator and facilitator of the learning process)	The Student— (In collaboration with teachers, family members, or dorm counselors)	The Parent or Guardian— (As the primary advocate for the student)
Theory to practice within a cultural, gender, racial, or linguistic context	considers the portfolio a mechanism to assist in melding theory, practice, professional belief, and cultural relevance;	writes or orally responds to a mathematics-development-and-attitudes survey using a cultural, gender, racial, or linguistic context;	assists the child with a response to a mathematics-development-and-attitudes survey, as well as makes cross-cultural comparisons;
NCTM's Standards	provides all stakeholders with a comprehensive plan for addressing the NCTM's Standards;	reviews and discusses the NCTM's Standards to be addressed during the academic year;	reviews and discusses with the child the NCTM's Standards and the cross-cultural comparisons to be addressed during the academic year;
Learning styles	integrates various aspects of teaching through learning styles;	develops awareness of learning styles (preferred styles, etc.);	provides background information to the child and the teacher concerning learning styles and developmental history;
Determines student mathematics-learning plan	collaborates with the student and parents to produce the student mathematics-learning plan and supports it;	organizes a mathematics-learning plan for students' improved achievement;	aids in developing a student mathematics-learning plan;
Collaboration	promotes collaborative experiences that rely heavily on problem-based learning with strong connections to native Life-Way teachings and survival skills.;	engages in collaborative experiences for problem-based learning with strong connections to native Life-Way teachings and survival skills;.	recognizes and encourages the learning behaviors for a successful group process required in problem-based learning, as well as native Life-Way teachings and survival skills;
Portfolio development	focuses on and openly addresses intellectual skills, not just behaviors, that are recognized by all students as legitimate academic skills;	develops a portfolio that gives evidence of mathematics achievement for continuous assessment, reflection, and goal setting;	supports and gives feedback for portfolio development, artifact selection, and setting;

	The Teacher— (As professional educator and facilitator of the learning process)	The Student— (In collaboration with teachers, family members, or dorm counselors)	The Parent or Guardian— (As the primary advocate for the student)
Assessment rubrics	develops and communicates to all stakeholders the mathematics-assessment rubrics that reflect the spirit of mastery learning;	understands and uses the mathematics-assessment rubrics developed for the mastery-learning process;	understands and communicates the mathematics-assessment rubrics used in the mastery-learning process;
Targeting growth	encourages students to use strengths while strengthening areas of weakness;	uses strengths while strengthening areas of weakness;	supplies additional evidence to identify the developmentally appropriate level for her or his child;
Formative and summative assessment	recognizes that the portfolio will be both a formative and summative assessment device to show growth over time;	recognizes that the portfolio can be a weekly, monthly, grading-period, and end-of-year assessment tool;	recognizes that the portfolio will be a weekly, monthly, grading-period, and end-of-year assessment tool to show growth over time;
Conferences as communication	uses the portfolio as the focus for assessment conferences;	engages in mathematics-assessment conferences using his or her portfolio;	engages in mathematics-assessment conferences with student and teacher;
Integration and transference of learning	appropriately integrates aspects of learning through varied curriculum designs and implementation processes;	shows evidence of knowledge transfer and connections among content areas;	encourages experiences that enhance the transfer of knowledge and connections for real-world learning;
Focus of assessment	assesses the student, not just the portfolio.	recognizes the importance both of artifacts as evidence and of a well-organized portfolio.	praises and supports the child's achievement, effort, and involvement.

Fig. 8.6. Mathematics portfolio-development matrix

Mathematics
Portfolio

Chei Ann Manygoats
1998—99

Verifine School

Gallup, N.M.

Student's Demographics

Prioritized Goals and Objectives

Annual Goals Based on NCTM's Standards

Short-Term Objectives

* Denotes student's own goal.

Goals		
Chei Ann	**DOB**	**Date**

NCTM STANDARDS

1. **Annual Goal:** To improve mathematics problem-solving skills by May 1999

Short-Term Objectives

* A. Chei Ann will continue to engage in problem-based-learning projects. She will be the group reader.

* B. Chei Ann will find and read at least three books that help with solving the problem.

C. Chei Ann will share all information with her group.

2. **Annual Goal:** To gain mastery over addition computational skills by May 1999

Short-Term Objectives:

* A. Chei Ann will be able to add numbers from 1 to 100.

B. Chei Ann will show 90 percent mastery for addition computations.

Self-Evaluation

**Chei Ann Manygoats
October 1998**

I can:
Add numbers from 1 to 100.

Add numbers without very many mistakes.

Find books to help with our group problem activities.

Question: How are you doing sharing your ideas with others?

Answer: I find the best books from the library to read to my group.

I need to give everyone time to share.

I like being the reader!

Fig. 8.7. An example of portfolio planning and development

- Will you distinguish between a display portfolio and a working portfolio?
- How will the artifacts be selected?
- Who will select them? Teacher? Student? Collaborative group?
- Will a rubric be used to guide assessment?
- How will the targeted NCTM Standards be selected?
- How will the portfolio be used in the total evaluation process (problem-based learning, discussion groups, reflections, collaborative projects, learning logs, etc.)?
- According to what criteria will the work in the portfolio be evaluated?
- How can the portfolio become a profile of the student—of his or her progress, needs, and culture?
- How can the portfolio lead to a gradual release of responsibility from teacher to student?
- How can the portfolio best include all educational stakeholders?
- Will you work individually or in teams when you evaluate the finished product?
- What kinds of information will you attempt to extract and summarize for institutional purposes? (from Yancey [1992])

(Inspired by Routman [1991] and Cleary and Peacock [1998])

Fig. 8.8. Questions to guide portfolio plans

- On what standards do you want to check progress in this unit of study? On this rubric?
- What will be your primary evidence to track progress on standards:
  — Authentic tasks in which students demonstrate significant outcomes?
  — Culminating exhibitions or project presentations?
  — Demonstration across the curriculum?
  — Reflective journals or learning logs?
  — Students' self-appraisal?
  — Oral interviews or conferences?
  — Checklists?
  — Review of products?
  — Engaging tests?
  — Students' observations?
  — Peer assessment?
  — Anecdotal records?
- Will you include a student's personal goals as outcomes to be assessed?
- How will you designate the level of accomplishment (e.g., Exceeds Expectations, Meets Expectations, Basic, Prebasic, Nonperformance, Other)?
- How will you set up the rubric
  — to be beneficial to you as a teacher?
  — to keep students positive, not defensive, about growing in skills?
  — to help parents understand their child's progress and the purpose of assessment?
- How does this rubric meet the assessment demands of the state, district, and school?

(Inspired by Cleary and Peacock [1998])

Fig. 8.9. Questions to guide preparation of outcome rubrics

REFERENCES

Abruscato, Joseph. "Early Results and Tentative Implications from the Vermont Portfolio Project." *Phi Delta Kappan* (February 1993): 474–77.

———. *Problem-Based Learning.* Alexandria, Va.: Association for Supervision and Curriculum Development, 1997.

Barrie, James. *Peter Pan.* New York: Charles Scribner's Sons, 1911, p. 79.

Carroll, Lewis. *Alice's Adventures in Wonderland.* London: Macmillan, 1866, 1872; New York: Grossett & Dunlap, 1946, p. 283.

Cleary, Linda M., and Thomas D. Peacock. *Collected Wisdom: American Indian Education.* Boston, Mass.: Allyn & Bacon, 1998.

Freire, Paulo. *Pedagogy of the Oppressed.* New York: Continuum, 1994.

Gilliland, Hap. *Teaching the American Indian.* 3rd ed. Dubuque, Iowa: Kendall/Hunt Publishing Co., 1995, p. 19.

Graves, Donald H., and Bonnie Sunstein, eds. *Portfolio Portraits.* Portsmouth, N.H.: Heinemann Educational Books, 1992.

Hiebert, Elfrieda H., ed. *Literacy for a Diverse Society: Perspectives, Practices, and Policies.* New York: Teachers College Press, 1991.

Joyce, Bruce, Martha Weil, and Emily Calhoun. *Models of Teaching.* 6th ed. Boston, Mass.: Allyn & Bacon Publishing Co., 2000.

Lazear, Daniel. *Multiple Intelligence Approaches to Assessment: Solving the Assessment Conundrum.* Tucson, Ariz.: Zephyr Press, 1994.

Lindsey, Randall, Kikanza Nuri Robins, and Raymond Terrell. *Cultural Proficiency: A Manual for School Leaders.* Thousand Oaks, Calif.: Corwin Press, 1999.

Lopez, Barry H. *Crossing Open Ground.* New York: Random House, 1989.

National Council of Teachers of Mathematics (NCTM). *Curriculum and Evaluations Standards for School Mathematics.* Reston, Va.: NCTM, 1989.

Quattromani, Libby, and Lyn Taylor. "Lesson Plans." In *The Encyclopedia of Mathematics Education,* edited by Louise Grinstein and Sally Lipsey. New York: Routledge & Falmer Publishing Co., 2001.

Ross, A. C. *Mitakuye Oyasin "We Are All Related."* 10th ed. Denver, Colo.: Wicóni Wasté, 1996.

Routman, Reggie. *Invitations: Changing as Teachers and Learners K–12.* Portsmouth, N.H.: Heinemann Educational Books, 1991.

Stokes, Sandra. "Curriculum for Native-American Students: Using Native-American Values." *Reading Teacher* 50 (April 1997).

Swisher, K., and D. Deyhle. "Adapting Instruction to Culture." In *Teaching American Indian Students,* edited by Jon Reyhner, pp. 81–95. Norman, Okla.: University of Oklahoma Press, 1992.

Taylor, Lyn. "An Elementary Teacher's Mathematical Life History." *Focus on Learning Problems in Mathematics* 18, nos. 1, 2, 3 (1996): 52–87.

Taylor, Lyn, and Libby Quattromani. "Integration of Elementary School Mathematics with Other Subject Areas." In *The Encyclopedia of Mathematics,* edited by Louise Grinstein and Sally Lipsey. New York: Routledge & Falmer Publishing Co., 2001.

Yancey, K. B. *Portfolios in the Writing Classroom.* Urbana, Ill.: National Council of Teachers of English, 1992.

Zemelman, Steven, Harvey Daniels, and Arthur A. Hyde. *Best Practice: New Standards for Teaching and Learning in America's Schools.* Portsmouth, N.H.: Heinemann Educational Books, 1993.

Student-Centered, Culturally Relevant Mathematics Activities

9

This article presents practical activities for teaching mathematics to American Indians, hereafter referred to as *Indians*. These activities can also serve to incorporate a cross-cultural component in non-Indian classrooms. The activities exemplify the spirit of the National Council of Teachers of Mathematics's *Curriculum and Evaluation Standards for School Mathematics (Curriculum and Evaluation Standards)* (NCTM 1989) and have been used successfully with Indian middle school students, including in summer projects that were held on the campus of the Colorado School of Mines by the National Science Foundation's Enhancement and Young Scholars programs for American Indians. These activities are also appropriate for use in elementary and secondary school mathematics classes.

NCTM's *Curriculum and Evaluation Standards* strongly encourages us to intensify our efforts to reach all students. American Indians constitute one of the groups that have not been reached often enough. The largest Indian subgroup in both reservation and urban communities is composed of children and adolescents (U.S. Bureau of Census 1984).

Indian students are at high risk educationally, in part because of the lack of relevance of their educational experiences (Williamson 1987). The school system, a reflection of the values of the dominant society, does not, for all practical purposes, take into consideration Indian values and traditional learning styles (Huffman, Sill, and Brokenleg 1986). "Bridging the gap between the Indian and non-Indian worlds is crucial to the success of schooling" (Little Soldier 1989). To reach Indian students, educators need to begin by using materials and teaching methods that are relevant to Indian cultures. Although many different tribes exist—the U. S. Department of Census currently recognizes more than 500 different tribes and 187 Indian languages (Trimble and Fleming 1989)—some common core values are shared across tribal boundaries (Locust 1988). Educators can use these values to enhance the participation of the Indian student in the learning process. Among the common values are cooperation, sharing, and harmony. (See Gilliland and Reyhner [1988] for a more complete listing and discussion of traditional values.)

Traditionally, Indian students come from extended families and tend to be group-oriented rather than self-oriented (Little Soldier 1989). Consequently, these students would tend not to respond well to competitive learning situations; they would likely prefer noncompetitive individual or cooperative

Lyn Taylor
Ellen Stevens

TEACHING ALL STUDENTS

The authors would like to thank Pam Boyer for her invaluable assistance with the preparation of this article.

learning activities. Little Soldier (1989) believes that "cooperative learning appears to improve student achievement, and it also matches such traditional Indian values and behaviors as respect for the individual, development of an internal locus of control, cooperation, sharing, and harmony" (p. 163). It can also improve students' attitudes toward themselves, others, and school, as well as increase "cross-racial sharing, understanding, and acceptance" (p. 163).

A person's attitude will likely affect both his or her motivation to pursue a task and subsequent achievement. One's mathematical attitude includes not only thinking and feeling but action (Taylor 1993). Therefore, this broad context suggests that a continual interaction occurs among a person's thoughts, feelings, and actions. This experience is unique for each individual because people bring a different history of life experiences to any given situation. As a result, the conceptual framework emphasizes the importance of providing learning situations in which students feel more comfortable and motivated to become active participants in their learning processes; this view is central to the NCTM's *Curriculum and Evaluation Standards* (1989).

The following exercises have been developed to engage students actively in learning and creating mathematics by providing a learning environment in which the student is an active participant. Since Indian students are non-competitive, these exercises can be used individually, with student pairs, or with cooperative groups. The following activities involve mathematical communication, pueblo modeling, tessellations, and finally, Pascal's triangle.

MATHEMATICAL COMMUNICATION

The NCTM's *Curriculum and Evaluation Standards* states not only that poor communication skills interfere with acquiring new knowledge but that teachers should be encouraged to include effective mathematics communication within their curriculum. The following activity, inspired by Freedman and Perl (1974), addresses this concern. This activity has been used successfully with students in the primary grades through high school, as well as with adults (Taylor 1991; Barone and Taylor 1996). The mathematical-communication activity involves students' working in pairs to learn communication skills in a sequential manner. An overhead transparency of a drawing, such as figure 9.1, is shown on the screen. One student is seated so that she or he can see the overhead-projector screen while the other is seated so that she or he cannot see it. The student facing the screen then describes to his or her partner how to draw the projected design. This activity can be used in a number of ways and can be adjusted to meet the diverse needs of students. The overall difficulty of the exercise can be altered by changing either the design presented to students or the pattern of students' interactions at increasingly difficult levels. As the levels increase in difficulty, so do the communication strategies that the students will be required to use.

- Level 1 (easiest)

  Through the use of two-way communication, the students doing the drawing are able to ask questions to clarify information. At this level, students are seated face to face and are able to see the drawing being done.

- Level 2 (somewhat harder)

  At this level, only the direction-giver may talk. Students remain seated face to face.

- Level 3 (more difficult)

  At this level, two-way communication between the students is still used; however, students remain seated back to back.

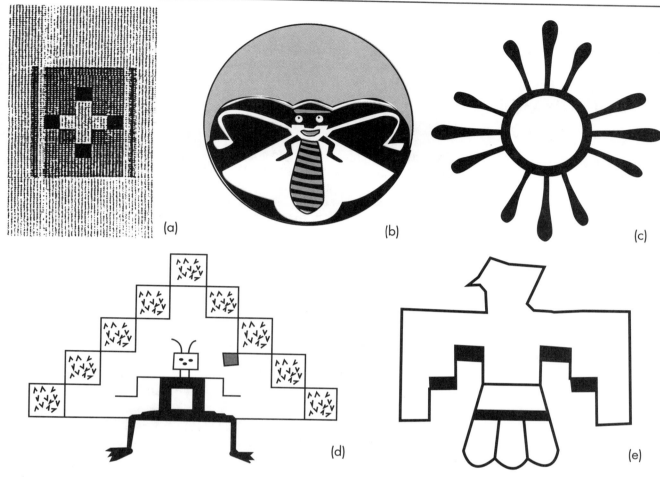

(a) Blackfoot design courtesy of Peregoy (1990); used with permission. Blackfoot bag and cradleband designs are usually rendered large with great masses of solid colors. These designs are found on pipe bags and in large pieces of beadwork.

(b) A depiction of the legend in which people and other creatures are lifted from the deepest of the four cave-wombs of the world; found on a Mimbres bowl. (Source: Fewkes 1928)

(c) The sun image is taken from petroglyphs in the Grand Canyon, Arizona. (Source: Ogintz 1994)

(d) This image is taken from Wenima (land of the dead) petroglyphs on a rock cliff in Hopi land (Source: Dutton and Olin 1991)

(e) The eagle design is from petroglyphs found in Bryce Canyon National Park, Utah. (Source: Ogintz 1994)

Fig. 9.1. Native American design for use on an overhead transparency in mathematical communication activity

- Level 4 (very difficult)

   At this level, only the direction-giver may talk; until the design is completed, the students remain seated back to back.

Manipulative materials can be used in an alternative version of this activity in which attribute blocks, pattern blocks, Cuisenaire rods, pretzels, pasta, or various cutout shapes replace the original projected drawing. With a divider placed between the students who are seated back to back (or side to side), one student is given a random selection of manipulative pieces with which to make a design, which student 1 then describes to the second student through two-way communication. Student 2 constructs the described design with her or his own set of manipulative materials. Encouraging the students to switch roles at all levels of the activity helps develop communication, listening, and visualization skills.

PUEBLO MODEL BUILDING

Creating three-dimensional scale models actively engages students in reasoning, estimating, measuring, and communicating. Teachers can use the school or other buildings with which the students are familiar to have them build scaled-down models. Having the students take actual measurements of these buildings gives them hands-on experience with measuring and making meaningful calculations by scaling down the actual measurements to create their models. (See Reif [1994] for descriptions of architectural mathematics in the classroom.)

An effective culturally appropriate use of this activity can be designed using *The Complete Guide to Building Taos Pueblo* (Jordon 1989), a paperback-book kit published by the Museum of New Mexico Press. The kit includes materials to build fifty-two authentic scale models of the Taos Pueblo in New Mexico, and an accompanying story about the pueblo. One kit can be used for an entire class or a large team of fifteen to twenty students by having each student construct two or three buildings. Students are responsible for constructing original printed buildings from the cut-out pieces included in the book and for building a second set that is proportionally scaled by a chosen factor. Figure 9.2 includes a picture of a pueblo built by Summer Math students.

Figure 9.2. A partially completed pueblo built by Lyn Taylor's 1995 Summer Math class

For this activity each group will need one *Building Taos Pueblo* kit; heavy paper, such as recycled manila file folders, for creating a second pueblo or scale model; scissors; glue; tape; measuring tapes or rulers; flat wooden toothpicks;

calculators; and pencils. The sheets of roofing materials included in *Hands-on Design Math* (Reif 1994) resemble adobe bricks and can be used for building scale-model pueblos. The grid markings on the materials make them especially useful for the scaling activities. The roofing materials have grid markings on the back, whereas the other sheets have grid markings on both sides to use for practice buildings.

After the teacher distributes a pueblo kit to the class, the students select a scaling factor to use to create a second pueblo model. Popular scaling factors include ×2 and ×3; more-challenging ones might be ×1/2 or ×3/2. Then students begin the building phase. Imagine all the problem solving, measuring, and other mathematics skills that the students use during this problem-centered activity. For example, small blocks or centimeter cubes can be used to calculate the volumes of various buildings before the structures are attached to their bases; students can be challenged to estimate the number of cubes that will fit in a given building. Each student can write an estimate on a sticky note with his or her name on it. Additionally, a student can make a graph of the class estimates on the chalkboard. After placing a few blocks in the building, students can be given the opportunity to change their estimate. The effects of the range, mean, and median of the data as students refine their estimates are interesting. When all fifty-two buildings are completed, they are assembled into a scale pueblo village on a base, contained on the inside cover of the book. The second scale pueblo model can be assembled on an appropriately sized poster-board or wooden platform. The dimensions were doubled in the second pueblo created by the Summer Math students, thereby quadrupling the surface area of the kit buildings, since doubling both the length and the width yields a surface area that is four times that of the original kit building. The volume of the second model building is eight times that of the original model, since doubling each of the length, width, and height yields a volume eight times larger.

Mathematical Communication and Integration Follow-up

This model-building activity can be extended in several ways. After completing the pueblo, students can write about their experience and the mathematics that they used during the activity. The *Building Taos Pueblo* (Jordon 1989) kit comes with twenty-four minipages of text, pictures, and a recipe for Pueblo bread. Each minipage is one-fourth of a book page and made to be removed and assembled into a smaller booklet to accompany the pueblo. This minibook could be used before the pueblo activity as an introductory social studies lesson. In addition to having students create their own factual books, a natural extension would involve their writing a story about the pueblo. Storytelling is very important to Pueblo people; see Livo (1985) for a further discussion of storytelling. "Becoming Mathematical Storytellers" was the theme of the 1996 Albuquerque Regional NCTM Conference and is becoming valued as an appropriate way of engaging students in mathematical communication.

Finally, making the Pueblo bread from the recipe included in the kit can lead to a hands-on-measurement experience that also integrates mathematics with home economics. Students can be challenged to double or halve the recipe or make one and one-half times as much. Students' understanding can be enhanced through their recognizing relationships among different mathematical topics and using mathematics in other curriculum areas, as well as in daily life (Jordon 1989, p. 32). The connections that we want students to explore should reflect authentic learning experiences rather than contrived "textbook" experiences.

INTEGRATING MATHEMATICS AND ART

Tessellations

Tessellations are mathematical exercises that allow for artistic creativity and geometric exploration. Geometry "provides an opportunity for students to experience the creative interplay between mathematics and art" (NCTM 1989, p. 158). The following are a suggested progression of culturally relevant activities that lead to more-advanced tessellations.

Tiling pattern blocks or attribute pieces can lead to some interesting mathematical discoveries and patterns (see Linquist [1989]). Figure 9.3 presents some tilings of regular polygons, that is, all the angles in the figure are congruent and all the sides are of the same length. Students from kindergarten through adulthood can enjoy and learn from this activity. Students can decorate their tilings with designs that are meaningful to them and their cultures.

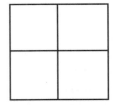

A four-square tiling

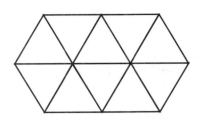

A tiling of ten equilateral triangles

Figure 9.3. Introductory tessellations

NCTM's *Curriculum and Evaluation Standards* (1989) suggests using the following activities for tiling tessellations (p. 158):

> For example, by repeated tracing of a regular polygon about a point so that the tracings coincide only along the edges and do not overlap, and then by extending the tracings outward on the paper (plane), students can discover whether the polygon might be used to form a tiling (tessellation) of the plane. Out of this very informal experience arises a fundamental question: Which regular polygons can be used to tile a plane?

Can you and your students create some tessellations? Students can experiment with pattern blocks or colored cardstock squares, equilateral triangles, hexagons, or other regular polygons. When used only one at a time, which regular polygons can be used to form a tiling? Which cannot? (Hint: adjacent sides need to meet at a common point; when three or more pieces come together, no gaps or spaces should appear at the intersection.)

Small groups and whole classes can discuss such questions as "What is a regular polygon?" and "Why can regular hexagons be tiled about a common point when regular pentagons cannot?" This discussion can lead more-advanced students to investigate how to determine the angle measure of regular polygons. (Example: lay four regular pentagons about a point. They overlap because the sum of the four angles that share a vertex is more than 360 degrees. The sum is 432 degrees. See Kaiser [1988] and Linquist [1989] for further information.)

Many extensions of this guided-discovery tiling activity can be explored. Students can experiment with tiling other quadrilaterals. They can discover that any quadrilateral can be tessellated across a plane because the sum of the interior angles is 360 degrees.

A more-advanced tiling activity involves cutting and pasting sections from a tessellating polygon; squares, rectangles, and rhombuses are good polygons with which to start. Figure 9.4 gives step-by-step directions for a sample problem.

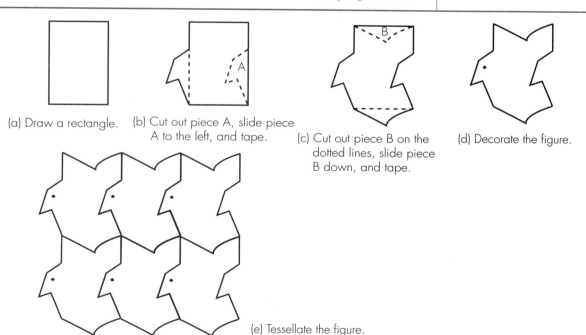

(a) Draw a rectangle. (b) Cut out piece A, slide piece A to the left, and tape. (c) Cut out piece B on the dotted lines, slide piece B down, and tape. (d) Decorate the figure.

(e) Tessellate the figure.

Note: Be cautious when creating your tessellating figure. Be sure to cut and slide the pieces directly. If you rotate or flip a piece before taping, the newly created figure may not tessellate.

Fig. 9.4. Creating a unique tessellation

Can students create their own tessellating figure? If so, can they tessellate it several times? An interesting follow-up activity, which is cooperative in nature, involves exchanging the newly created tessellation with another person or with a team of two persons. Students can then further extend and tessellate one another's designs. A cardboard pattern or graph paper can be helpful aids in this pursuit. Figure 9.5 is an example of a graph-paper tessellation.

Congruent triangles of any shape can be used for a tessellation. They can be placed side by side to cover a surface without any overlapping and without leaving any spaces uncovered.

Step 1: Finish covering the grid with additional congruent triangles of the same size. Count squares carefully to see that each new triangle drawn is congruent to those already shown.

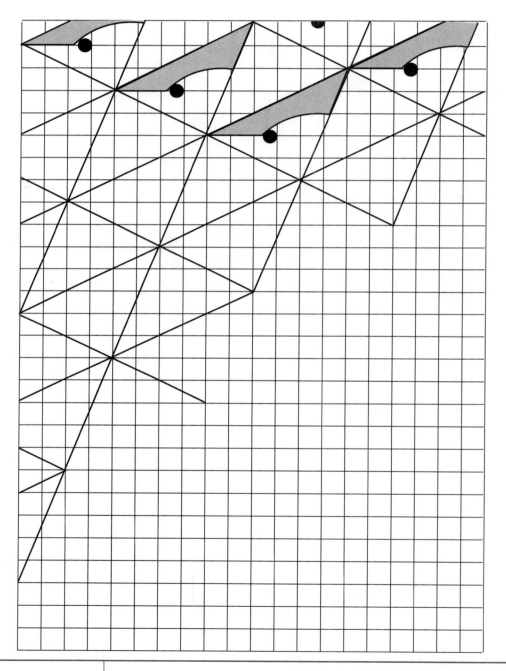

Fig. 9.5. Tessellating with triangles (Source: Maletsky [1974, p. 336])

Clo Mingo's students at Santa Fe Indian School created the tilings in figures 9.4, 9.6, 9.7, and 9.8. Their tessellations could be decorated using symbols from their cultures. Can students identify the original polygon from which each tessellated figure was made? Can they extend the tessellation by repeating the design? Have students pick one of the tessellations and try to extend it. Have students discuss with another person how they extended the tessellation.

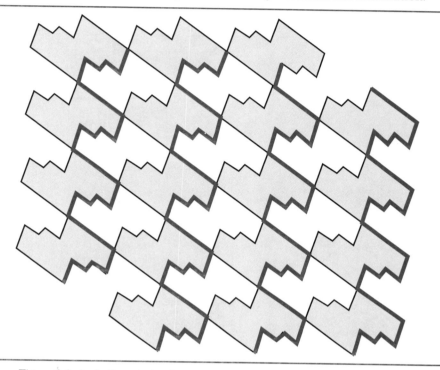

Figure 9.6. An Indian student's creative tessellation achieved using a rhombus

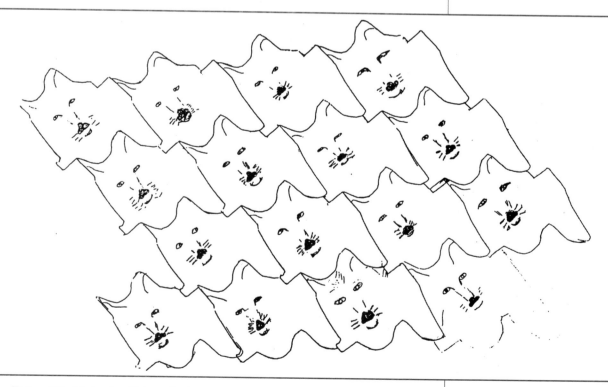

Figure 9.7. A lobo (wolf) tessellation created by a Santa Fe Indian student

Advanced tessellation activities could include looking for examples of tessellations and other geometric patterns and symmetries in Indian artwork, such as sand paintings, pottery, beadwork, baskets, rugs, and jewelry. The tessellated eagle in figure 9.8 was created by a very artistic Indian student. This design is culturally relevant, as the eagle holds a special place of honor in Indian cultures. It is the highest-flying bird and is thought to speak to the Great Spirit. The eagle also represents a spiritual ally and a source of power across many Indian tribes (Doore 1988).

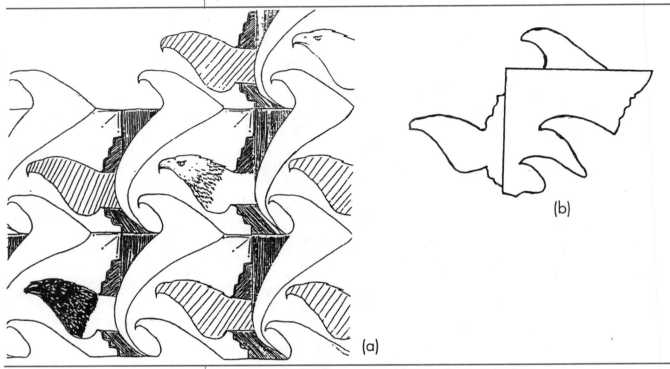

Figure 9.8. (a) Advanced tessellation and (b) the cardstock pattern from which the eagle tessellation was created

Pascal's Triangle

Another activity relevant to the Indian culture involves Pascal's triangle. Figure 9.9 presents several rows of Pascal's triangle and step-by-step "leading questions" to assist teachers and students in developing additional rows and discovering some of the patterns. How many rows does Pascal's triangle have? The illustration in figure 9.9 shows a specific number of rows, but the triangle can always be extended by adding another row. In each row of the triangle, can students find a pattern that leads to a shortcut for completing the row? Have them discuss the symmetric row pattern with a friend. For further information, readers may want to explore some books that discuss the patterns and usefulness of this special triangle (e.g., _Pascal's Triangle_, by Green and Hamburg [1986]; _Visual Patterns_, by Seymour [1986]; _Discovering Geometry: An Inductive Approach_, by Serra [1989]).

Repeatedly embedding Pascal's triangle in an Indian beadwork design can be the basis for a creative worksheet for a culturally relevant activity. Such a design, developed by Vera Preston, was created by Ramona Mason, an artist with the Region V Native American Resource Center in Tulsa, Oklahoma (Skinner 1985). Figure 9.10 illustrates this colored beadwork representation

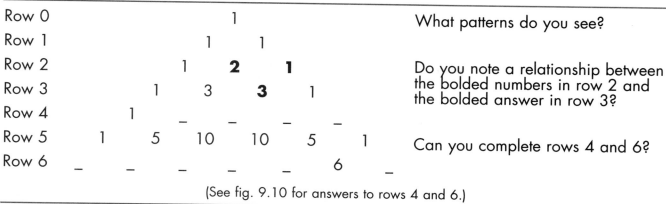

Row 0						1						
Row 1					1		1					
Row 2				1		**2**		**1**				
Row 3			1		3		**3**		1			
Row 4		1		–		–		–		–		
Row 5	1		5		10		10		5		1	
Row 6	–		–		–		–		–		6	–

What patterns do you see?

Do you note a relationship between the bolded numbers in row 2 and the bolded answer in row 3?

Can you complete rows 4 and 6?

(See fig. 9.10 for answers to rows 4 and 6.)

Figure 9.9. Pascal's triangle

1. Yellow	4. Turquoise	7. Dark green	20. White
2. Black	5. Grey	10. Pink	21. Red
3. Green	6. Orange	15. Purple	35. Blue

Figure 9.10. Pascal activity

using Pascal's triangle. Each number is assigned a color. Although many colors can be selected, the colors used in figure 9.10 are significant for some Indian people because they symbolically represent stages and attitudes in life. For example, yellow can depict youth, the sun, or west; green signifies growth or east; black indicates introspection or south; and white represents north and the hardship of the winter snow, which develops strength. Teachers and students can cooperatively create other beadwork patterns using Pascal's triangle. Can students tessellate Pascal's triangle?

Readers who want even greater challenge can examine the mathematics and history of M. C. Escher's tessellations (see Escher [1970]; Haak [1976]; Ranucci [1974]; Zurstadt [1984]).

Indian artwork and craftwork are as varied as the number of tribes. One of the common factors in beadwork is a reliance on geometric patterns; patterns can also be found in other forms of Indian artwork (e.g., rugs, baskets, and pottery). This observation does not suggest that all Indians do beadwork or weaving but rather points out that geometry is a shared element in these art forms (see Bradley [1975]). Bradley (1992, 1993) also uses Logo to create a Navajo blanket design and a Sioux directions design.

Symmetry and Patterns

Symmetry and geometric patterns can be explored using a Mira with designs from books or student-created designs. A Mira can also be used to create the other half of a student-created totem pole. See figure 9.11. Totem poles are typically carved by North American Indians from the northwestern sections of the United States and western Canada. Additional examples of patterns found on pottery, clothing, and the like are shown in *NCTM Student Math Notes* (Barkley 1997). Math of the Navajo (see fig. 9.12), a Key Curriculum Press poster (Swienciki 1992), illustrates the importance spatial visualization, geometry, and the mathematics of Navajo weaving.

Can you draw the other half of this totem pole?

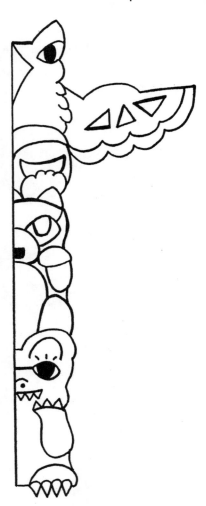

Fig. 9.11. A student-created totem-pole worksheet

Math of the Navajo, Key Curriculum Press, 1150 65th Street, Emeryville, CA 94608, 1-800-995-MATH, www.keypress.com; used with permission

Fig. 9.12. Math of the Navajo poster

Connecting Mathematics with Literature

One of our favorite examples of connecting mathematics with literature uses *The Popcorn Book* (dePaola 1978), a story about the history of popcorn and its use among American Indians. The author tells us that "popcorn was discovered by the Indian people in the Americas many thousands of years ago" (p. 10). He also writes that "archeologists found some popped corn that was 5600 years old" in a "bat cave in New Mexico" (p. 11). This book creates the environment for a popcorn-based estimation activity that incorporates various concepts from NCTM's *Curriculum and Evaluation Standards* (1989) in one activity. An estimation jar can be filled with popcorn kernels or popped popcorn. Since popcorn kernels are fairly small, students find that this activity is more challenging than just estimating the number of pieces of popped popcorn. Estimating the weight of the kernels and that of popped popcorn are other challenging activities. Using an old-style popcorn maker with the lid off and placed in the center of a large plastic sheet, students can predict how far the popped corn will fly when it is popped and then actually measure where the popcorn landed. Graphing the results can also be interesting for elementary school students.

The Popcorn Book (1978) can be used to introduce a study of the author dePaola. *The Legend of the Indian Paintbrush* (dePaola 1988) is an Indian story that was retold to the author. Other literature books that can be used to create integrated lessons include *Turquoise Boy, Navajo Legend* (Cohlene 1990), *Iktomi and the Boulder, a Plains Indian Story* (Gobel 1988), and *Children of the Earth and Sky* (Krensky 1991). See Welchman-Tischler (1992) and Thiessen, Matthias, and Smith (1992) for further ideas on using children's literature to teach mathematics.

CONCLUSION

We have presented information concerning Indian values, as well as several detailed activities that can be used to help bridge the gap between mathematics and our diverse students. The mathematical content and methodologies incorporated in the activities presented here are compatible with the NCTM's *Curriculum and Evaluation Standards* (1989). Students are actively involved in communicating mathematically, exploring geometric and numerical patterns, connecting mathematics with other subjects, and relating mathematics to life and culture. The information and activities presented in this article can be effectively used by thoughtful teachers to reach more students—both Indian and non-Indian.

The success of our experiences, including the development of positive mathematical attitudes (see Taylor and Stevens [1988]), may be attributable to the noncompetitive and cooperative nature of these activities. The culturally relevant experiences appeared to facilitate the development of positive mathematical attitudes, thereby affecting students' thinking, feeling, and actions. We hope that experiences such as these will encourage Indian students, as well as others, to continue with their mathematics education and to view mathematics as meaningful, connected with other subjects, and of interest and importance in their lives. We also hope that this article has given the reader some ideas to assist in opening up the beautiful world of mathematics to reach more students.

BIBLIOGRAPHY

Barkley, Cathy. "Native American Patterns." *NCTM Student Math Notes* (September 1997): 1–4.

Barone, Michelle, and Lyn Taylor. "Peer Teaching with Math Manipulatives: A Practical Guide for Teachers." *Teaching Children Mathematics* 3 (September 1996): 8–15.

Bradley, Claudette. "Native American Beadwork Can Teach Mathematics." Unpublished manuscript, Harvard University Library, 1975.

———. "Teaching Mathematics with Technology: Creating Sioux Four Directions with LOGO." *Arithmetic Teacher* 39 (November 1992): 46–49.

———. "Teaching Mathematics with Technology: Making a Navajo Blanket Design with LOGO." *Arithmetic Teacher* 40 (November 1993): 45–50.

Cohlene, Terri. *Turquoise Boy, a Navajo Legend.* Mahwah, N.J.: Watermill Press, 1990.

dePaola, Tomie. *The Popcorn Book.* New York: Scholastic Book Services, 1978.

———. *The Legend of the Indian Paintbrush.* New York: G. P. Putnam's Sons, 1988.

Doore, Gary, ed. *Shaman's Path: Healing, Personal Growth and Empowerment.* Boston: Shambrala Publishing, 1988.

Dutton, B., and C. Olin. *Myths and Legends of the Indians of the Southwest.* Santa Barbara, Calif.: Bellerophon Books, 1991.

Escher, M. C. *The Graphic Work of M. C. Escher.* New York: Hawthorn Books, 1970.

Fewkes, J. Walter. "Designs on Prehistoric Pottery from the Mimbres Valley, New Mexico." *Smithsonian Miscellaneous Collections* 74 (1928): n.p.

Freedman, Miriam K., and Teri Perl. *A Sourcebook for Substitutes.* Menlo Park, Calif.: Addison-Wesley Publishing Co., 1974.

Fries, J. E. *The American Indian in Higher Education: 1975–1976 to 1984–1985.* Washington, D. C.: U.S. Government Printing Office, 1987.

Giganti, Paul, Jr., and Mary Jo Cittadino. "The Art of Tessellation." *Arithmetic Teacher* 37 (March 1990): 6–16.

Gilliland, Hap, and John Reyhner. *Teaching the Native American.* Dubuque, Iowa: Kendall/Hunt Publishing Co., 1988.

Green, Thomas M., and Charles L. Hamberg. *Pascal's Triangle.* Palo Alto, Calif.: Dale Seymour Publications, 1986.

Gobel, Paul. *Iktomi and the Boulder, a Plains Indian Story.* New York: Orchard Books, 1988.

Haak, Sheila. "Transformation Geometry and the Artwork of M. C. Escher." *Mathematics Teacher* 69 (December 1976): 647–52.

Huffman, Terry E., Maurice L. Sill, and Martin Broken Leg. "College Achievement among Sioux and White South Dakota Students." *Journal of American Indian Education* 25 (January 1986): 32–38.

Hunt, W. Ben, and J. F. Buck Burshears. *American Indian Beadwork.* New York: Macmillan Publishing Co., 1951.

Jordon, Louann C. *The Complete Guide to Building Taos Pueblo.* Santa Fe, N.M.: Museum of New Mexico Press, 1989.

Kaiser, Barbara. "Explorations with Tessellating Polygons." *Arithmetic Teacher* 36 (December 1988): 19–24.

Krensky, Stephen. *Children of the Earth and Sky.* New York: Scholastic, 1991.

Lane, Melissa. *Women and Minorities in Science and Engineering.* Washington, D.C.: National Science Foundation, 1988.

Lindquist, Mary M. "Implementing the Standards: The Measurement Standards." *Arithmetic Teacher* 37 (October 1989): 161–63.

Little Soldier, Lee. "Cooperative Learning and the Native American Student." *Phi Delta Kappan* 71 (October 1989): 61–63.

Livo, Norma. *Storytelling.* Denver, Colo.: Libraries Unlimited, 1985.

Locust, Carol L. "Wounding the Spirit." *Harvard Educational Review* 58 (March 1988): 315–30.

Maletsky, Evan M. "Activities: Designs with Tessellations." *Mathematics Teacher* 67 (April 1975): 335–38.

Martin, Hope. "Readers' Dialogue: Tessellations and History." *Arithmetic Teacher* 37 (February 1990): 2, 8.

National Council of Teachers of Mathematics (NCTM). *Curriculum and Evaluation Standards for School Mathematics.* Reston, Va.: NCTM, 1989.

Ogintz, Eileen. *Taking the Kids to the Great American Southwest.* New York: HarperCollins Publishers, 1994.

Peregoy, John. Personal interview, 1990.

Ranucci, Ernest R. "Master of Tessellations: M. C. Escher, 1898–1972." *Mathematics Teacher* 67 (April 1974): 299–306.

Reif, Daniel K. *Hands-on Design Math.* Palo Alto, Calif.: Dale Seymour Publications, 1994.

Seattle Public Schools, Mathematics Office. *Multicultural Mathematics Posters and Materials.* Reston, Va.: National Council of Teachers of Mathematics, 1983.

Serra, Michael. *Discovering Geometry, an Inductive Approach.* Berkeley, Calif.: Key Curriculum Press, 1989.

Seymour, Dale. *Visual Patterns in Pascal's Triangle.* Palo Alto, Calif.: Dale Seymour Publications, 1986.

Skinner, Linda. *Integration of Knowledge: Culturally Related Learning Activities, Book 2.* Oklahoma City: Oklahoma State Department of Education, 1985.

Stanley-Millner, Pamela. *Authentic American Indian Beadwork and How to Do It.* New York: Dover Publications, 1984.

Swienciki, Lawrence. "Math of the Navajo." Berkeley, Calif.: Key Curriculum Press, 1992. Poster.

Taylor, Lyn. "A Phenomenological Study of Female and Male University Professors' Life Histories in and Attitudes toward Mathematics." Ph.D. diss., University of New Mexico, 1988.

———. Unpublished classroom activities, 1991.

———. "Vygotskian Influences in Mathematics Education, with Particular Reference to Attitude Development." *Focus on Learning Problems in Mathematics* 15 (spring/summer 1993): 3–17.

———. "Activities Integrating Mathematics and Culture." In *Multicultural and Gender Equity in the Mathematics Classroom: The Gift of Diversity,* 1997 Yearbook of the National Council of Teachers of Mathematics (NCTM). Reston, Va.: NCTM, 1997.

Taylor, Lyn, and Ellen Stevens. *American Indian Science and Engineering Society Summer Math Camps Evaluation Report.* Final Report to American Indian Science and Engineering Society, 1988.

Taylor, Lyn, Ellen Stevens, John J. Peregoy, and Barbara Bath. "American Indians, Mathematical Attitudes, and the Standards." *Arithmetic Teacher* 38 (February 1991): 14–21.

Thiessen, Diane, Margaret Matthias, and Jacquelin Smith. *The Wonderful World of Mathematics.* Reston, Va.: National Council of Teachers of Mathematics, 1997.

Trimble, Joe J. E., and Candice C. M. Fleming. "Providing Counseling Services for Native American Indians: Client, Counselor, and Community Characteristics." In *Counseling across Cultures,* 3rd ed., edited by Paul Pedersen, Juris Draguns, Walter Lonner, and Joseph Tremble, pp. 177–204. Honolulu: University of Hawaii Press, 1989.

U.S. Bureau of the Census. *Census Fact Sheet: A Statistical Profile of the American Indian Population: 1980 Census.* Washington, D. C.: U.S. Government Printing Office, 1984.

Welchman-Tischler, Rosamond. *How to Use Children's Literature to Teach Mathematics.* Reston, Va.: National Council of Teachers of Mathematics, 1992.

Williamson, Karla J. "Consequence of Schooling: Cultural Discontinuity." *Canadian Journal of Native Education* 14 (spring 1987): 60–69.

Zurstadt, Betty K. "Tessellations and the Art of M. C. Escher." *Arithmetic Teacher* 31 (January 1984): 54–55.

Weaving a Multicultural Story

10

Clo Mingo

> Actually, weaving is an ancient art in this country, dating back several thousand years. Woven baskets were used to store and cook food, to hold water, and to carry objects; they even served as cradles and hats.
> —Anna Lee Walters, *The Spirit of Native America*

Whether intended for embellishment or for practical purposes, weaving—of fiber, vines, animal hide, feathers, reeds, bark, cloth, corn husks, paper—was and is a part of almost every culture's folk art. Woven baskets have been widely used as containers, cooking utensils, and water vessels. Floor mats and whole dwellings have sometimes been woven from the materials at hand. Hats are woven in Africa, Asia, and Central America. Navajo weavers make beautiful rugs to hang, use as horse blankets, or spread on the floor.

To incorporate Native American students' experiences, culture, and philosophy into the mathematics curriculum, high school mathematics classes at the Santa Fe Indian School scheduled six or seven special projects during each semester. This school operates under contract to the Bureau of Indian Affairs in Santa Fe, New Mexico, and is not part of any school district or county school system. Differences in available materials, previous mathematical and artistic experience, and personal dexterity account for striking similarities and differences among the projects created by students from different Native American cultures. The imagination, creativity, geometry, and weaving skills that emerged from one of these efforts, the Woven Hearts Special Project, was extraordinary. In the project, Native American students adapted a Norwegian folk art (*julekurv*) and made it their own.

THE PROJECT

The students were given the basic instructions on the sheet shown in figure 10.1. Students worked in groups during class and completed the assignment individually as homework that was due in ten days. The preliminary activity conducted in class familiarized them with the project.

From a mathematical standpoint, the students had looked at the heart shape only as an example of symmetry, and they called the familiar curved heart the "symmetrical heart" (see fig. 10.2). Although woven hearts can be made using this familiar shape, its lack of straight edges makes for warp strips that are uneven and difficult to work with. By contrast, using what students came to call the "geometric heart" allowed them to use intricate weaving designs and gave them an entirely new way to look at the two-dimensional heart shape.

THE PREVIEW

In the group activity that introduced the Woven Hearts Special Project, many students were encountering and using tools and terms that were new to them.

Woven Hearts Special Project

Name _____

Basic Woven Heart

1. Using a square and two semicircles, make a heart pattern on card stock. (The diameter of the circle has to be the same as the sides of the square.)

2. Using your pattern, cut two hearts out of paper of two different colors.

3. Cut the square parts of your hearts into equal strips.

4. Be creative and original in the design and decoration of your woven heart.

♥ ♥

Self Evaluation

1. The grade I deserve on this project is _____ because _____
   _____
   _____

2. What I learned from this project was _____
   _____
   _____

3. What I enjoyed most about this project was _____
   _____
   _____

4. What I liked least about this project was _____
   _____

Prizes will be offered for the smallest heart and the best-decorated heart.

Fig. 10.1. The instruction and self-evaluation sheet for the Woven Hearts Special Project

Mathematically, students have only looked at the heart shape as an example of symmetry, but although woven hearts can be made using what the students called the symmetrical heart, the lack of straight edges makes warp strips that are very uneven and difficult to work with.

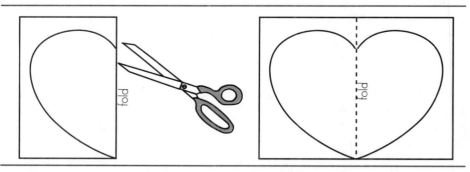

Fig. 10.2. An example of a "symmetrical heart"

For the activity, students were seated at tables in groups of three to five. The activity lasted fifteen to twenty minutes and required an example of a two-color woven heart. In addition, the following supplies were necessary:

- Colored scratch paper
- Scissors
- Metersticks
- Tape
- Plastic right triangles

- Protractors
- Compasses

High school students have often forgotten terminology that was introduced in previous years, so the activity served as a review of relevant vocabulary, including the terms *side, diameter, radius, right angle, metric units,* and *congruent.*

Students were allowed sufficient time—fifteen to twenty minutes—so that their first attempts at constructing the geometric heart would be successful experiences that would generate interest in the assignment. After using a compass to produce a circle that could be divided into two semicircles, the students used a protractor to make a square with a side equal to the circle's diameter (see fig. 10.3). Some students were surprised to discover that tracing around a protractor did not necessarily produce a semicircle; usually another student could explain that the midpoint on the straightedge of the plastic protractor was not necessarily the center of the circle and would point out the hole that was the midpoint of the diameter.

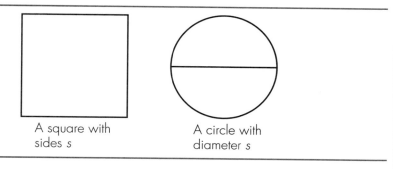

A square with sides *s*

A circle with diameter *s*

Fig.10.3

The students then cut and taped the two semicircles to adjacent sides of the square and centered the figure on the point opposite the intersection of the circles. Alternatively, some students constructed a geometric heart on a single piece of paper by drawing a square using a plastic right triangle and then constructing semicircles on two adjacent sides of the square with a compass (see fig. 10.4).

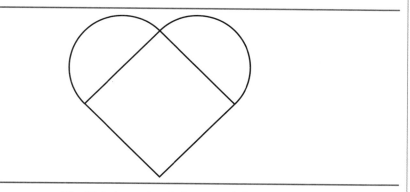

Fig. 10.4. A constructed geometric heart

Students quickly devised methods for constructing the geometric heart without using the standard tools of geometry, but instead using techniques that showed their excellent understanding of the relevant geometric concepts. They discovered that drawing around the bottom of a coffee cup makes a fine circle that they could cut out and fold in half to produce two semicircles. They realized that they could easily make a square from a rectangular sheet of paper by folding a corner against the opposite side and trimming away the paper that extends beyond the folded isosceles right triangle.

Students then measured the square part of the geometric heart and divided it into four equal warp strips (see fig. 10.5). They discovered that measuring and cutting accurately and carefully made the weaving much easier later. Next, with a second color, they measured and cut four equal weft strips (see fig. 10.5). The simplest version of weaving would use paper of one color for both the warp (vertical) strips and the weft (horizontal) strips. Requiring the students to use two colors immediately introduced one variation of the plain weave.

The students then wove the weft strips alternately over and under the warp strips. They learned to press or slide each row of weaving with their fingers so that the strips just touched each other. The students were left on their own to be as creative, resourceful, and precise as they could be.

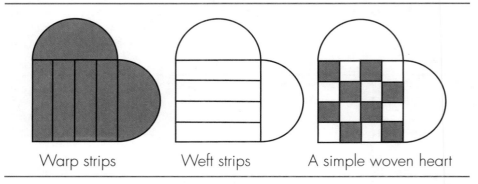

| Warp strips | Weft strips | A simple woven heart |

Fig. 10.5

THE RATIONALE

Because the Woven Hearts Special Project was the sixth or seventh special project out of fourteen scheduled throughout the school year to augment the regular high school mathematics class work, the students were familiar with the general objectives:

- Incorporating other curriculum areas into mathematics
- Exposing students to the role of women and minorities in the field of mathematics
- Demonstrating practical uses of mathematics
- Rewarding creativity
- Providing opportunities to plan ahead
- Using technology
- Encouraging cooperative learning
- Assessing by self-evaluation

A schedule of special projects for two semesters is shown in figure 10.6. Each objective contributed a unique facet to the Woven Hearts Special Project.

Incorporating Other Curriculum Areas into Mathematics

Because the woven heart is a warmly remembered artifact from my own Norwegian American childhood, I told the students about my own experiences of weaving the hearts from red, green, and white construction paper as two-dimensional Christmas tree decorations. I also told them about weaving three-dimensional May baskets to fill with flowers and hang on doorknobs. We talked about the fact that some cultures use the heart shape in connection with Christmas and May Day instead of just for Valentine's Day, as in American

---

Special Projects—Year II

First Semester

September 11	Summer Math *
September 25	Scale Drawings *
October 16	Origami
November 6	Biography of a Mathematician *
	(one male and one female)
November 20	Tessellations *
December 11	Ojo de Dios
January 15	Snowflakes *

Second Semester

February 5	Woven Hearts (Julekurv)
February 19	Math in Advertising *
March 11	Garden Project *
March 25	Kites
April 15	1040EZ *
May 6	Egg Project *
May 20	Research Project

*Computers or calculators must be used in some way.

Fig. 10.6

---

culture. We also discussed occupations that use the heart shape, such as baking, floral design, the arts, and the like.

Exposing Students to the Roles of Women and Minorities in Mathematics

By focusing on weaving, the Woven Hearts Special Project introduced and showed the mathematical dimensions of an occupation that many cultures have tended to regard as "women's work." Although both men and women have been and are weavers in various cultures, any craft, skill, or occupation that tends to be gender-specific to females is usually excluded from textbooks and the traditional school experience. This omission reinforces the idea that such an activity is of less value than activities customarily associated with males. It is therefore very important to explore mathematical patterns and accomplishments of females and minorities in building the mathematical skills of all members of the class.

Demonstrating Practical Uses of Mathematics

In classes of Native American students who represent several different tribal affiliations, teachers can expect to have students whose families have extensive experience in rug weaving, basket weaving, and beadwork. Some students may already be expert craftspeople, whereas others will have no experience with traditional crafts. A discussion with my students about using mathematics in planning a design, ensuring symmetry, and estimating amounts of various necessary materials provided additional background for the Woven Hearts Special Project while showing mathematics in a practical light.

Metric measuring tools were used in the project, as they are in all projects where such tools are applicable, since many high school students have great difficulty transferring their knowledge of the customary measurements that appear in most American textbooks to the metric measurements used in their science classes. Using the proper metric measuring tools in tasks like the Woven Hearts Special Project helps familiarize students with the metric system.

Rewarding Creativity

The Woven Hearts Special Project, like all special projects, emphasized the importance of being creative instead of simply fulfilling the assignment. Because creativity received particular attention, it was also recognized and rewarded. Each special project was discussed and displayed before the individual class, and whenever possible, the work was displayed in the classroom or hallways for at least a day or two. Displaying mathematical activities outside the classroom heightens interest in mathematics, develops appreciation for creativity, helps recruit students for the next year's classes, and brightens the school environment.

Providing Opportunities to Plan Ahead

The Woven Hearts Special Project followed the pattern of the other special projects in giving the students two weeks between the assigning of the project as homework and its due date. The two-week notice for these projects is intended to give students opportunities to plan ahead—to schedule their homework around sports, work, dates, and family fun and responsibilities—as well as to emphasize the importance of assignments that cannot be completed in one evening.

Using Technology

Although technology in mathematics is often regarded narrowly as consisting of computers and calculators, projects like the Woven Hearts Special Project can show students that traditional, "low-tech" tools of geometry (compass, protractor, meterstick, triangle, etc.) are also important. A short video clip of several different weavers at work familiarized students with the operation of a loom and served to motivate them for the Woven Hearts Special Project.

Encouraging Cooperative Learning

The introductory class activity involving woven hearts was designed as a cooperative activity that allowed students who were less skilled in following instructions, using tools, measuring correctly, and visualizing a finished product to learn from those who were more skilled in these areas. As they worked together, the students brainstormed about variations that they might try, and these interactions served as useful catalysts for creativity. I was tempted to show examples of particularly imaginative woven hearts, but doing so might have inhibited students' creativity, since students often find it easier to copy someone else's idea than to create their own designs.

Assessing by Self-Evaluation

Self-evaluation (see fig. 10.1) is an integral part of all the special projects. Students may at first be uncomfortable and insecure about evaluating themselves, but practice helps them understand that it will be important throughout their lives to have a goal in mind and to ask themselves whether they have achieved it. The self-evaluation in the Woven Hearts Special Project, as in the others, elicited from the students why they thought they had earned a particular grade and encouraged them to supply reasons that were based on their personal assessments of their talents, creativity, and understanding of the assignment. Every student was required to fill in every part of the self-evaluation form.

> From the earliest times, Native Americans maintained a unique aesthetic vision of the universe and never was the vision separated from the functional aspects of their culture.
>
> —Anna Lee Walters, *The Spirit of Native America*

THE PRODUCT

Students' products in the Woven Hearts Special Project exceeded all expectations, going far beyond the project's goals of using the tools of geometry, emphasizing accuracy in measurement, and encouraging creativity. The woven hearts that the students produced ran the gamut of weaving styles, reflecting their artistic skills, imagination, and personal exposure to Native American weaving designs.

Most students submitted woven hearts that employed some variation of plain weave but with many more warp and weft strips than the four in the example. Unknowingly, the students discovered for themselves the three basic weaving patterns—plain, basket, and twill—and they created classic variations, which afforded opportunities for investigation, categorization, identification, and discussion—all skills necessary for mathematics classes. Some of the students' creations are shown in figures 10.7 to 10.12.

The heart in figure 10.7 appeared to present a classic basket weave (described by Wilson [1971, p. 33]), with two wefts of blue over two warps of black. Cutting the warp and weft strips on a diagonal, however (at a 45° angle to the semicircle instead of at the usual 90° angle), made a much more difficult and complicated variation that produced the patterns that formed the outside borders of the weaving.

Fig. 10.7. A basket weave on the diagonal by a Santa Fe Indian School student

One student developed overshot pattern areas (floats) by covering groups of warp strips with weft strips. Although the design (fig. 10.8) appeared to be random, three predominant green weft strips wove a pattern with three white warp strips. The floating white warp strips made this overshot weave unusual.

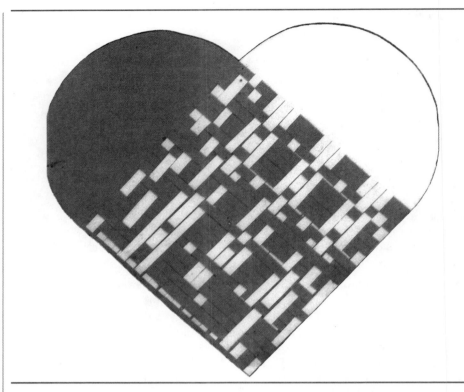

Fig, 10.8. An overshot weave by a Santa Fe Indian School student

Another student developed a plain-weave variation (fig. 10.9) by cutting with pinking shears both the warp strips (which were hot pink) and the weft strips (which were black). By carefully weaving a weft strip of one color over a warp strip of another, the student had created a heart that appeared to contain reflected, stepped pyramids that had been tessellated. (The students had learned the terminology to describe this variation in a previous special project.)

Fig.10.9. A plain weave with pinked edges by a Santa Fe Indian School student

Three other students' variations were also noteworthy. The first student produced a plain-weave variation by using black warp strips and white weft strips to provide a dramatic background for tiny Native American symbols drawn on a random selection of weft squares (fig. 10.10). The second student found that decreasing the widths of the warp and weft strips gave the plain weave an interesting op-art effect (fig. 10.11; see Wilson [1971, p. 74]). The third student's heart exhibited a complex, slightly asymmetric example of the diamond twill pattern in black and blue, with Native American symbols on the nonwoven parts of the heart.

Fig. 10.10. A plain weave with Native American symbols by a Santa Fe Indian School student

Students sometimes pasted colored semicircles over white ones to create more colorful, pleasing designs. The woven hearts shown in figures 10.7, 10.11, and 10.12 show this process.

Some students created designs that turned out to have names, although these were unfamiliar to us. The school library was a useful resource for books and videotapes that proved to be very useful and informative about weaving, and we learned new patterns and names from them.

THE ASSESSMENT

In their views of the spiritual and physical world, the tribes use symbols freely to enrich their daily lives and ceremonies.

—Anna Lee Walters, *The Spirit of Native America*

The students' evaluations of their own work were very enlightening. For example, it turned out that the student who had designed the basket weave on the diagonal (fig. 10.7) was dissatisfied with his work because he had wanted all

Fig. 10.11. A plain weave with perspective by a Santa Fe student

Fig. 10.12. A diamond twill weave with Native American symbols by a Santa Fe Indian School student

the borders to be the same. His classmates were very supportive but tried to persuade him that this goal might not have been attainable on the "inside" borders. The students advanced fascinating arguments about the capacity of even or odd numbers of pairs of warp and weft threads to produce "finished" or "raw" edges. Several students reached eagerly for graph paper to illustrate their ideas.

The Pueblo student who had used the overshot weave in figure 10.8 wrote that he should get a C+ for his work. In his view, he "didn't get it right," even though the heart he had submitted was his eighth try. He had wanted to cross-stitch designs that Pueblo women used on gingham aprons. He underestimated the grade value of his projects, as many students do when they have to put in writing why they think they deserve a certain grade. Successful self-evaluation takes practice and honesty, but even first attempts at self-evaluation can be revealing to both students and teachers.

One girl, hoping to win the prize for the smallest woven heart, stapled to the top of the form a precisely woven heart that was 0.5 cm across. She wrote that her goal had been to make it so small that the teacher would have to put on her reading glasses in order to see it. She succeeded!

It turned out that the design of the heart in figure 10.9 had been influenced by constraints imposed by the flu. Confined to bed, the student could find only a pair of pinking shears and some scraps of poster board, but she sent the woven heart that she produced with these materials to school with a classmate so that it could be turned in on time.

The boy who submitted the heart in figure 10.10 admitted that he had planned to draw a small design in each of the white squares but had run out of time. Because he had not accomplished his goal, he gave himself a B on the assignment, even though he had ultimately decided that the randomly interspersed white spaces probably made for a better, simpler design.

While we were discussing this heart, the students raised a question about designs that different cultures have in common. They were interested to learn that there is good evidence that almost all cultures use the same basic geometric designs. From antiquity to the present day, these patterns have been used for decoration on pots, bodies, fabrics, stones, and baskets (Kellogg, Knoll, and Kugler 1965, pp. 1129–30). The students were intrigued to discover that these images have a name, *phosphenes*, which describes the patterns of dots, checkerboards, grids, circles, crosses, spirals, parallel lines, and so on, that people see when their eyes are closed.

The student who had designed the op-art heart (fig. 10.11) had been studying perspective in his art class and had been inspired to play with the idea. His fascination with the mathematical applications of perspective continued for the remainder of the year.

All the classes voted the heart in figure 10.12 as the best-decorated woven heart. The student who had created the heart then turned it over to show its original front. Her first designs had not worked well at all, but she had been so pleased with the woven design, which had taken her so long to do, that she had decorated the back instead.

The term "phosphene" refers to the images perceived by the human brain as visual images in the absence of visual stimuli.... Phosphenes can result from a variety of causes, including gentle pressure on closed eyes, migraine headaches, fasting, etc. (Patterson 1992, p. 155)

EXTENSIONS

I have successfully used a number of extensions of the Woven Hearts Special Project:

- Making three-dimensional heart baskets. *Julekurv* also denotes a three-dimensional heart-shaped basket constructed of paper. Filled with cookies and candies, it is a traditional Norwegian Christmas decoration symbolizing hospitality (Kenny 1985, pp. 128–29). As figure 10.13 shows, the three-dimensional baskets are made with folded (or doubled) half hearts. The first time I assigned the Woven Hearts Special Project, I could not remember how to do the three-dimensional weaving, so I challenged the students to figure it out. This tactic proved to be so useful that I've conveniently "forgotten" how to make the three-dimensional heart (or Christmas basket) ever since. Both high school students and participants in preservice and inservice teacher workshops have risen to the challenge. I highly recommend the ploy. (I consider the task of making the baskets an excellent topology problem, but it is really easy once you have the two sections in your hands.)

- Simulating woven hearts on the computer. Making and decorating a woven heart that has been simulated with any computer drawing program challenges students to be creative with technology. I have found that designing a Navajo rug using The Geometer's Sketchpad (1991) can be a very effective way to introduce students to the diverse tools and possibilities of this software, which has been used effectively to demonstrate some remarkable designs similar to those traditionally found in Navajo rugs.

- Examining additional examples of weaving done by people in other cultures. Bringing a Navajo rug into the classroom, for example, gives students an opportunity to identify several mathematical elements and concepts: parallel lines, perpendicular lines, circles, squares, rectangles, chevrons, symmetry, reflection, rotation, and translation.

- Studying weaving from the perspectives of other disciplines. Collaborating with teachers in art, language arts, and social studies can provide avenues for pursuing other aspects of, or ideas about, weaving that students have brought up. For example, my students became interested in other types of weaving, designs that were common across several societies and cultures, and stories about other countries' traditions—particularly May Day, a holiday that they do not celebrate. Other teachers in other fields could help them explore these ideas. A consideration of phosphenes in science class was also a possibility.

REFLECTION

Despite a fairly common approach to daily life, tribal societies were often diverse; their languages, dwellings, clothing, utensils, and crafts differed widely, even when tribes shared a particular environment or came into close contact with others.

—Anna Lee Walters, *The Spirit of Native America*

Native American tribes share many common elements in their philosophies, traditional lifestyles, ceremonies, and crafts. However, with more than three hundred tribes in the United States alone, there are vast differences among traditional tribes and considerable variation even within tribes because of increased mobility, intermarriage, and the impact of the media. The luxury of a monocultural classroom is quickly disappearing in all areas of our society.

As a teacher, I have conducted research in, taught graduate courses about, and developed several archaeomathematics units that are based on Native

American (Anasazi) mathematics in Chaco Canyon, New Mexico (Mingo 1997). I have incorporated technology and collaborative learning into all my work while continually trying to refine the curriculum by listening to students talk about their lives, their dreams, and their cultures. Sometimes I have succeeded, and sometimes I have failed, but the special projects have inspired me and my students to incorporate multicultural goals and to think creatively in the mathematics classroom.

Professional Standards for Teaching Mathematics (see the quote at the right [NCTM 1991, p. 146]) challenges mathematics educators to look at curriculum, pedagogy, and students in a new way. Weaving a multicultural story through the Woven Hearts Special Project is one way to put mathematical concepts in context and to incorporate students' experiences, language, viewpoints, and culture.

Teachers also need to understand the importance of context as it relates to students' interest and experience. Instruction should incorporate real-world contexts and children's experiences and, when possible, should use children's language, viewpoints, and culture. Children need to learn how mathematics applies to everyday life and how mathematics relates to other curriculum areas as well.

REFERENCES

The Geometer's Sketchpad. Berkeley, Calif.: Key Curriculum Press, 1991.

Kellogg, Rhoda, Max Knoll, and Johann Kugler. "Form Similarity between Phosphenes of Adults and Pre-School Children's Scribblings." *Nature* 208 (1965): 1129–30.

Kenny, Maxine. *Folk Art Christmas Ornaments: How to Make Them.* New York: Arco Publishing, 1985.

Mingo, Clo. "Grounded Practice: Lessons in Anasazi Mathematics Emerging from the Multicultural Classroom." In *Multicultural and Gender Equity in the Mathematics Classroom: The Gift of Diversity,* 1997 Yearbook of the National Council of Teachers of Mathematics (NCTM), edited by Janet Trentacosta, pp. 177–85. Reston, Va.: NCTM, 1997.

National Council of Teachers of Mathematics (NCTM). *Professional Standards for Teaching Mathematics.* Reston, Va.: NCTM, 1991.

Patterson, Alex. *A Field Guide to Rock Art Symbols of the Greater Southwest.* Boulder, Colo.: Johnson Books, 1992.

Saltzman, Ellen Lewis. *Overshot Weaving.* New York: Van Nostrand Reinhold, 1983.

Wilson, Jean. *Weaving Is Fun.* New York: Van Nostrand Reinhold, 1971.

Using Native American Legends to Explore Mathematics

11

In recent years, numerous resources have been published describing the use of literature to aid in the study of elementary school mathematics (Burns 1992; Welchman-Tischler 1992; Whitin and Wilde 1992, 1995; Zaslavsky 1996). In fact, *Teaching Children Mathematics* has often featured articles that link literature and mathematics. Literature has the potential to engage students in mathematics in a nonthreatening manner, as well as to provide a real context in which to situate mathematical study.

Generic stories of all types contain mathematical connections, and many stories contain both a mathematical and a multicultural connection. The regular use of such resources permits multicultural connections to occur throughout the school year rather than be relegated to specific times during the year. Further, they demonstrate to children that all cultures find uses for mathematics.

This article presents specific activities for use with a Cheyenne legend about the creation of the Big Dipper, one of the familiar constellations in the northern hemisphere. The activities are appropriate for use with upper elementary or middle school students and provide an avenue for integrating mathematics with social studies, language arts, and science.

In the legend of the Big Dipper (see Goble [1988]), a young Cheyenne girl begins to make beautiful shirts and moccasins for seven brothers whom she has never met. After completing all the outfits, she sets off on a trip to find her brothers. Once she locates them, she becomes an integral part of their family and joins in their daily activities. One day, the family is threatened by a herd of stampeding buffalo. To save the family, the youngest brother shoots arrows into the sky, causing a tree to grow along the path of the arrows. The young girl and her brothers climb the tree to safety. As the lower branches of the tree are pounded by the buffalo and begin to break, the youngest brother continues to shoot arrows higher and higher, causing the tree to grow taller and taller. Eventually, the young girl and her brothers reach the region of the stars and become the Big Dipper.

This Cheyenne legend can lead to exploring and studying rich ideas in the mathematics classroom. One such idea may involve the study of geometry, as Native American art is highly geometric in nature. Another idea addresses the topic of measurement, offering students ample opportunities to link topics within mathematics and to connect mathematics with other disciplines. General

Denisse R. Thompson
Michaele F. Chappell
Richard A. Austin

SUMMARY OF *HER SEVEN BROTHERS*

The authors wish to thank Stefanie Dishong and her fifth-grade students at Riverhills Christian School, Tampa, Florida, for their assistance with these activities.

descriptions follow of sample activities that emphasize certain geometry and measurement topics. These activities can be implemented in various instructional formats, such as cooperative groups, learning-center tasks, or whole-class discussion. Subsequent to the descriptions, we share our experiences using these activities with children.

MATHEMATICS ACTIVITIES

Activity Sheet 1: A Designer of Clothes

By design, Native American art is often quite geometric. Activity sheet 1 (pp. 137–39) provides a real context in which students can study geometry patterns, particularly symmetry, and numerical or algebraic ways of describing those patterns. To simulate the use of dyed porcupine quills, students can use colored toothpicks to create designs with specific types of symmetry, then describe their patterns. By then posting the designs on a bulletin-board display, students have many opportunities to consider different geometric designs that contain a given number of lines of symmetry.

Question 4 offers a natural extension to the other activities in this lesson. If time and conditions warrant, students can explore their own environment to determine what natural items might be useful in creating designs. Different items might be available in the school and home environments, providing opportunities for comparison and contrast.

Activity Sheet 2: Searching for Her Brothers

The second activity sheet (pp. 140–41) gives students a real context in which to explore measurement; refer to questions 1 and 2. In fact, the activities described on this sheet represent a natural opportunity to integrate mathematics with social studies. As students explore the mathematics, they can simultaneously explore the history of the Plains Indians. Question 3 affords an opportunity to compare the area occupied by the Cheyenne during the 1800s with the area occupied by their reservations today. The use of ratios to make this comparison arises from a real context. Question 4 presents an opportunity to examine the rate at which buffalo were brought to the brink of extinction, as well as to assess the impact of the loss of buffalo on the lives of the Plains Indians.

Activity Sheet 3: Reaching the Stars

The final activity sheet (pp. 142–43) presents an opportunity to integrate mathematical concepts of measurement with science concepts, specifically, heights of trees and distances to the stars. Unlike computations on the previous activity sheets, these computations require students to have access to a scientific calculator. The questions on the activity sheet relate to the tallest tree on Earth and the closest star outside our solar system. A natural extension is to have students consider the height of a typical tree in their location (or the distance to the Sun or their favorite planet) and determine the number of times the arrow would need to be shot to reach that height (or distance).

"Reaching the Stars" also presents a good opportunity for students to work in small cooperative groups. In particular, different groups of students could investigate the legends of different star constellations and then report their findings to the class.

"A Designer of Clothes"

The activities described above are appropriate for use with upper elementary or middle school students. For example, fifth-grade students became quite involved with "A Designer of Clothes." After reading the story to the class, the teacher informally introduced the concept of lines of symmetry by having students express any previously learned ideas about this concept. The students described a line of symmetry as the place where one could cut a figure in half so that both parts would be identical.   With this introduction, the students were shown several pages from the book and were asked to identify the lines of symmetry in the outfits made by the young girl for her brothers.

After this discussion, each student was given a Mira with which to explore aspects of symmetry. The students began by determining the lines of symmetry on the blank rug template of question 2. Students were also encouraged to think about alphabet letters that have just one or two lines of symmetry, to write those letters, and to check their predictions with the Mira.

Once students had a basic understanding of lines of symmetry, we gave each student several toothpicks of different colors and asked them to create rug designs with one, two, and no lines of symmetry, then to check their designs with the Mira. Despite the previous discussion about symmetry and the students' responses, which suggested understanding on their part, the actual design of rugs proved to be difficult for some of the fifth graders. In particular, many students initially created designs with two lines of symmetry rather than one; hence, teachers might want to consider whether switching the order of questions 2a and 2b would be better for their students.

Figures 11.1–11.3 illustrate some of the designs that the students created. The student who created the design in figure 11.1 initially had at the lower edge of her design four toothpicks that were identical in color to those at the upper edge. After some discussion with the teacher, she realized that she needed to have different colors at the upper and lower edge to avoid having a horizontal line of symmetry in addition to the vertical line of symmetry. Figures 11.2 and 11.3 illustrate two other straightforward examples. Although students were not as creative in their designs as we might have desired, they were successful with this task and seemed to enjoy the activity.

Fig.11.1. Victoria's rug design with exactly one line of symmetry

Fig. 11.2. Marsha's rug design with exactly two lines of symmetry

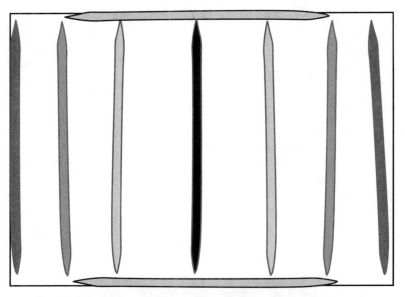

Fig. 11. 3. Brian's rug design with exactly two lines of symmetry

After completing designs with one, two, and no lines of symmetry, students were asked to consider making a rug with four lines of symmetry. For this example, we did not provide any template, yet students began to realize that a square rug was needed. The students' previous use of the Mira proved particularly helpful for this task because they had already determined that the diagonals in a rectangle were not lines of symmetry. Although we focused on the concept of lines of symmetry, teachers could extend this lesson by having students design rugs with turn symmetry of various magnitudes.

Question 3 affords an opportunity for students to explore patterns, an essential aspect of informal algebraic thinking. As a class, we considered one fifth-grade student's rug pattern and counted the number of toothpicks of each color that the student had used in making the design. We then determined the number of each color needed to make two rugs, three rugs, seven rugs, and one hundred rugs, giving most of the students a good opportunity to use mental mathematics. After completing the table, we attempted to have students write a general rule that would yield the number of toothpicks needed regardless of the number of rugs made. Although the students could verbalize the rule, our introducing the letter *r* for the number of rugs initially proved to be confusing for the fifth graders. However, with further discussion and examples, the students appeared to understand the application of the rule for the pattern.

The students were then asked to complete the table in question 3a for one of their own rug designs and then write a rule for a friend, as suggested in question 3b. Although most of the students could complete the table without difficulty, many had problems trying to write a rule for their designs. Figures 11.4–11.6 illustrate some of the rules that the students wrote. The rule in figure 11.4 is fairly clear; the example supplied is interesting in that the student's design used eight red toothpicks rather than five. The rule in figure 11.5 is reasonably clear as written in words; however, the algebraic rule contains the number 100, which has no connection with the values in the table. Finally, the rule in figure 11.6 gives the number of each color in just one rug. Even though several of the students had difficulty in writing the abstract rule for their designs, informal experiences such as these are important precursors to further work with algebra and variables.

b. Suppose that a friend wanted to copy your design to make some rugs. Write a rule that would tell your friend how many of each color of toothpick would be needed.

Example:
I wanted to
make 3 rugs
and 5 toothpicks
for each rug, I multiply
$3 \times 5 = 15$ toothpicks
for 3 rugs.

Rule: Multiply your number of red
toothpicks to the number of rugs
you plan to make.

Fig. 11.4. Michelle's rule for the number of toothpicks for her rug design

b. Suppose that a friend wanted to copy your design to make some rugs. Write a rule that would tell your friend how many of each color of toothpick would be needed.

example
$(100 \times R)$

multiply the number of
tooth picks by the number
of rugs

Fig. 11.5. Ashley's rule for the number of toothpicks for her rug design

b. Suppose that a friend wanted to copy your design to make some rugs. Write a rule that would tell your friend how many of each color of toothpick would be needed.

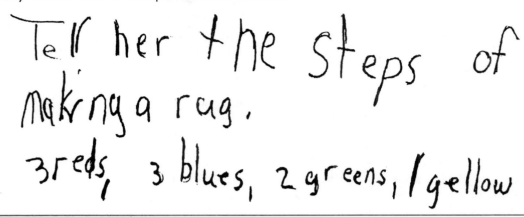

Tell her the steps of making a rug.

3 reds, 3 blues, 2 greens, 1 gellow

Fig. 11.6. Donald's rule for the number of toothpicks for his rug design

Searching for Her Brothers

Questions 1 and 2 on activity sheet 2 encourage students to explore measurement concepts. We visited the same fifth-grade classroom about a week after completing the activities on sheet 1. The students summarized the story and reviewed the concepts of line symmetry before beginning this new lesson. As a class, we completed the table for question 1a through the entry for twenty days. When students reached the item requiring the number of miles walked in a month, they raised the question about the number of days in a month. Students were permitted to choose their favorite month, write it on the line, and then determine the appropriate distance walked. As an extension and to build on the work with patterns and rules from the first visit, we asked the students to write a rule giving the number of miles traveled for any number of days. More students were successful at writing this rule than had succeeded in writing the rule for the number of toothpicks in the rug. Figure 11.7 contains one student's table and rule.

a. Suppose she walked 10 miles each day. How far would she walk in two days? Three days? Seven days? Twenty days? A month?

Number of Days	Number of Miles Walked
1 day	10 miles
2 days	20 miles
3 days	30 miles
7 days	70 miles
20 days	200 miles
1 month December	310 miles
One year	3650 miles

b. Suppose she needed to walk 250 miles each day. How many days would it take her to walk this far? How many weeks is this?

> The number miles you walk you times it by the number of days When you times it that is the number of miles you walked.

Fig. 11.7. Victoria's table and rule for the number of miles walked

Elementary school students often lack a good feel for given distances. Before beginning question 2, the students were asked to walk a distance that they thought was 100 feet. Some students carefully counted their paces; others, confusing yards and feet, raced to a distance that they thought represented the length of a football field; and still others made wild guesses. Using tape measures, the students then measured a distance of 100 feet. With a stopwatch, we measured the time needed by each student to walk this distance. Returning to the classroom, students used calculators to determine their own results to questions 2a and 2b.

This particular group of fifth graders, having not had much experience with this type of problem solving, needed considerable guidance through the process of determining their time to walk a mile. We began by having students multiply their time, in seconds, to walk 100 feet by 52.80, which is the number of 100-foot groups in a mile. Because the number of seconds was quite large, students then rewrote their time in seconds in more familiar units. Natural extensions to this task could involve having students use their walking time to compute the length of time needed to walk from home to school or the time needed to walk the length or breadth of a park in their neighborhood.

We touched only briefly on the area issues raised in question 3. The students used a grid to cover the portion of the Great Plains identified as the territory of the Cheyenne. By counting the number of square units of the grid covering the region, students began to explore area concepts from a conceptual point of view.

Reaching the Stars

Activity sheet 3 was used in an interview setting with a sixth grader. The table in question 1a is quite similar to the table in question 1a on activity sheet 2. The task posed no difficulty for this sixth grader; in fact, she quite easily wrote a general rule for the pattern. However, the science investigations beginning in question 2, which resulted in the introduction of scientific notation, seemed challenging. Although this concept is appropriate for middle school mathematics and science students, this sixth-grade student had not yet been exposed to such notation. Because the computations were completed with a calculator, it became necessary to discuss how scientific notation is recorded on a calculator display. The experience suggests that such an activity as the one described here affords teachers a natural opportunity to introduce this topic. With the increased use of technology, such topics as scientific notation are increasingly important to the middle school curriculum, and we must find realistic means of introducing them.

CONCLUSION

Literature is one motivating resource for exploring mathematics. By using literature resources with a multicultural context, teachers can engage students in exploring the mathematics used by many cultures. Multicultural connections then become a natural theme in the classroom rather than an add-on feature studied only occasionally. This article presents one such resource, furnishing for upper elementary and middle school students a series of activities related to a particular Native American legend.

REFERENCES

Burns, Marilyn. *Math and Literature (K–3): Book One.* Sausalito, Calif.: Math Solutions Publications, 1992.

Goble, Paul. *Her Seven Brothers.* New York: Aladdin Books, 1988.

Welchman-Tischler, Rosamond. *How to Use Children's Literature to Teach Mathematics.* Reston, Va.: National Council of Teachers of Mathematics, 1992.

Whitin, David J., and Sandra Wilde. *Read Any Good Math Lately? Children's Books for Mathematical Learning, K–6.* Portsmouth, N.H.: Heinemann, 1992.

———. *It's the Story That Counts: More Children's Books for Mathematical Learning, K–6.* Portsmouth, N.H.: Heinemann, 1995.

Zaslavsky, Claudia. *The Multicultural Math Classroom: Bringing In the World.* Portsmouth, N.H.: Heinemann, 1996.

Activity Sheet 1: A Designer of Clothes

Read the story *Her Seven Brothers* by Paul Goble (1988). Then answer the questions below.

1. Look at the designs of the shirts and moccasins that the young girl made for her brothers. Describe the symmetry that you see in each outfit.

2. Suppose that the young girl decided to make a rug for each brother to use in the tepee. Use a set of colored toothpicks to create a rectangular rug meeting each of the following conditions.

   a.   A rug with exactly one line of symmetry

b.   A rug with exactly two lines of symmetry

c.   A rug with no lines of symmetry

3. Look back at the designs that you made above. Select one of your rug designs, and count the number of each color of toothpick that you used to make the rug.

   a.   How many toothpicks of each color would you need to make two rugs? Three rugs? Four rugs? Seven rugs? One hundred rugs?

Number of Rugs	Color: _____ Number of Toothpicks	Color: _____ Number of Toothpicks	Color: _____ Number of Toothpicks
1 rug			
2 rugs			
3 rugs			
7 rugs			
100 rugs			

b. Suppose that a friend wanted to copy your design to make some rugs. Write a rule that would tell your friend how many of each color of toothpick would be needed.

4. In the story, the young girl used items from nature, specifically porcupine quills, to decorate the outfits that she made. Take a walk outside your classroom or your house. What items do you find that you could use to decorate a shirt? With your classmates or with the members of your group, collect a sufficient quantity of items, then use them to design a new shirt. Describe any symmetry in your final design.

Activity Sheet 2: Searching for Her Brothers

1. The young girl in the story walked for many days until she found the tepee of her brothers.

   a. Suppose that she walked 10 miles each day. How far would she walk in two days? Three days? Seven days? Twenty days? A month?

Number of Days	Number of Miles Walked
1 day	
2 days	
3 days	
7 days	
20 days	
1 month	

   b. Suppose that she needed to walk 250 miles. How many days would it take her to walk this far? How many weeks is this number of days?

   c. Suppose that the young girl in the story can walk a mile in 30 minutes. How long would it take her to walk 10 miles? 250 miles?

2. Use a ruler to measure a distance of 100 feet. Use a timer, such as a stopwatch, to find how long it takes you to walk this distance. Try to walk at a pace that you would be able to walk all day long.

   a. At that pace, how long would it take you to walk 1 mile?

b. At that pace, how long would it take you to walk 10 miles? To walk 250 miles?

3. The Cheyenne were Great Plains Indians who moved around in search of buffalo.

a. On a map of the United States, find the region of the country where the Cheyenne lived. Put a grid over this region, and estimate the area of the region over which the Cheyenne roamed looking for buffalo.

b. Find out some information about the size and location of reservations on which many Cheyenne live today. Compare the size of the reservations with the size of the region in which they lived in the past.

4. In the 1800s, many buffalo roamed the Great Plains. Find out some information about the number of buffalo during that time, some of the uses for buffalo, and the number of buffalo today. Write a short report sharing your findings. Use tables and charts as needed.

Activity Sheet 3: Reaching the Stars

1. In the story, the youngest brother saved his sister and other brothers from the buffalo by shooting an arrow into the sky. A tree grew along the path of the arrow. His sister and older brothers then climbed the tree to safety. When they got to the top of the tree, the youngest brother shot another arrow, and the tree grew higher. He continued this process until his sister and brothers reached the stars.

   a. Suppose that each time the boy shoots an arrow, it travels 75 feet. How tall would the tree grow if he shot the arrow 2 times? 3 times? 5 times? 100 times?

Number of Arrows Shot	Height of Tree
1 shot	
2 shots	
3 shots	
5 shots	
100 shots	

   b. Among the tallest trees in the world are the California redwoods. Find out the typical height reached by these trees. How many times would the young boy need to shoot the arrow to have his tree grow to this height? (Assume that his arrow travels 75 feet each time.)

2. Light travels at a speed of 186,000 miles per second.

   a. Find out how far light travels in a day, a week, and a year.

   b. The distance that light travels in a year is called a *light-year*. Suppose that the young boy could shoot his arrow a distance of 1000 feet each time. How many times would he need to shoot the arrow to cover a distance of one light-year?

3. The closest star to Earth outside our solar system is Alpha Centauri, which is approximately four light-years away.

    a.   How many times would the boy need to shoot the arrow to reach Alpha Centauri, assuming that each time his arrow travels 1000 feet?

    b.   How long do you think it would take the boy to reach Alpha Centauri? Justify your answer.

4. Find out about other legends that tell the story of the creation of different constellations of stars. Share your findings with other members of your class.

Kakaanaq

A Yup'ik Eskimo Game

12

Esther A. Ilutsik

Games in many different cultural and ethnic groups reflect aspects of the cultural traditions during a certain time period. For example, many Arctic indigenous groups have very similar games of physical strength and endurance. They are currently found in such events as the "Eskimo Olympics" and include such amusements as the ear pull, the seal hop, and the high kick. These games no longer serve the traditional purposes of gaining physical strength and testing endurance that were valuable for the traditional hunting and gathering lifestyles of the Arctic indigenous people. The games now serve to remind Arctic natives of their heritage.

Although games associated with physical strength and endurance continue to be passed down, very little attention has been given to other games that also include improving skills required for the hunt. One of the most important things to learn for the hunt is to look at an object, judge the distance, and determine how much thrust to apply to an object in one's hand to hit the object accurately. Developing this skill takes a lot of practice and patience.

The Yup'ik people of western Alaska created a game to develop this skill that young boys learned and played into manhood. The Yup'iks call this enjoyable and challenging game *kakaanaq*. This lesson introduces the strategy of kakaanaq and emphasizes different mathematical concepts that are integral parts of the game.

KAKAANAQ

Kakaanaq was explained to me in 1997 by Elder Henry Alakayak Sr. of Manokotak, Alaska, who grew up in the since-abandoned village of Kulukak. Kulukak is located in the western part of Alaska, on the southern coast of the Bering Sea, in the Bristol Bay region, where the indigenous people are predominantly Central Yup'iks. The Bristol Bay region is a mountain-bordered basin in the southwest corner of Alaska, with the Kuskokwim Mountains to the west and north and the Aleutian range to the south and east.

Elder Henry Alakayak Sr. recalls watching and also participating in kakaanaq as a young boy and a young man in the village of Kulukak. The purpose of the game was to increase the skills required in hunting—patience, visual acuity, aim, and accuracy in judging the thrust required for throwing objects of different weights. Kakaanaq can be played at any level but is most suitable for elementary school children.

Materials

Traditionally, the "target mat" was a piece of animal skin placed on the ground. The target mat measured approximately twenty-four inches by twenty-four inches with a small circular dot in the middle that measured about one and one-half inches in diameter. Each of two players had five circular disks that measured three inches in diameter (see fig. 12.1). These disks were made of wood, with each set of disks bearing different drawings (possibly to identify the players).

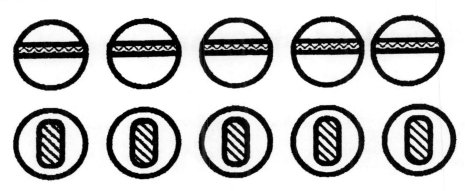

Fig. 12.1. Examples of the disks used in kakaanaq

For classroom use, the target mat can be made from poster board or a material that has a semirough surface, such as a piece of fabric. The target in the center of the mat can be drawn with a colored permanent marker (see fig. 12.2). The disks can be either handmade or purchased from a local crafts store or a teacher's supply store or catalog. They can be colored or decorated to distinguish the players' pieces.

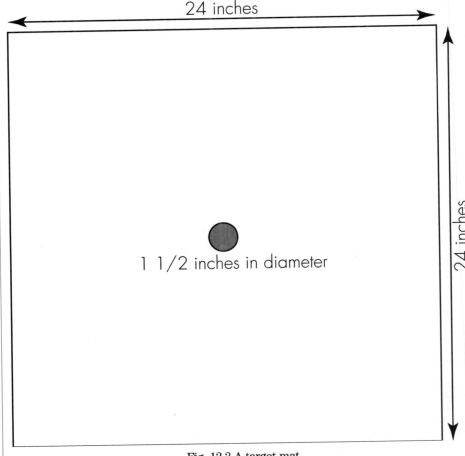

Fig. 12.2 A target mat

PLAYING KAKAANAQ

Before a kakaanaq match began, the players would traditionally determine what the winner would receive. The loser might do one of the winner's chores, such as getting wood for the *maqi*—a Yup'ik term for *sauna*—or give the winner an item that was desired by either player. Then the players would decide where they would stand in relation to the target mat. They would begin three feet from the target mat and would increase the distance as their skill improved. The players would decide who would begin first. With each of them equipped with five disks, they would take turns attempting to hit the target.

The object of the game was to land the disks on the target. If a disk covered the entire target, the player would score 3 points. If it covered half the target, the player would score 2 points. If the disk merely touched the target, the player would score 1 point (see fig. 12.3). Players could score points for themselves or their opponents (see fig. 12.4) and could also change the score in their favor by sliding their opponents' disks off the target. If they were successful in these attempts, any points that their opponents had earned for their now-displaced disks would be deducted from their scores (see fig.12.5). The game continued until one player reached 10 points.

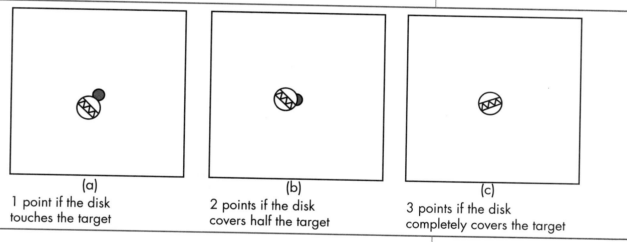

(a)	(b)	(c)
1 point if the disk touches the target	2 points if the disk covers half the target	3 points if the disk completely covers the target

Fig. 12.3. Scoring points earned

If player A ( ) slides player B's ( ) disk so that it touches or covers the target, player B receives points as shown.

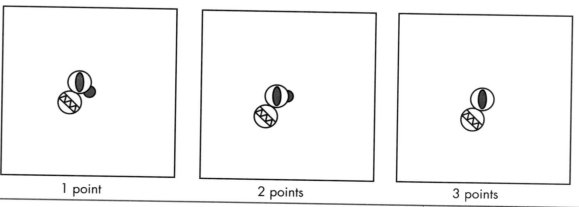

1 point	2 points	3 points

Fig. 12.4. Scoring points for the opponent

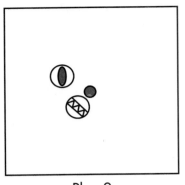

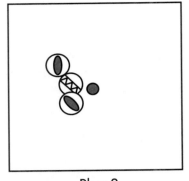

Play 2   Play 3   Play 4

Points can be lost if a player slides his or her opponent's disk off the target.

For example, if player A ( ) and player B ( ) are playing a game and player A scores one point, the target mat would look like the one shown in figure 12.3a. In play 2, player B scores 1 point and player A loses the point earned in her first turn. In play 3, player B loses 1 point and player A does not score because the disk does not touch the target. In play 4, player A does not gain any points, but player B scores 2 points. The score at the end of the fourth play is 0 (player A) to 2 (player B).

Fig. 12.5. Scoring points lost

To play kakaanaq in the classroom, teachers can begin by having the players determine what good deed or simple task the loser will do for the winner. The good deed might be helping carry the winner's books or helping with a simple chore that the teacher might ordinarily assign to students (e.g., erasing the board). The players should place the target mat on the floor against a wall and measure three feet from the nearest edge of the mat, using masking tape to mark the distance. Players should take turns tossing the disks in the customary fashion.

To introduce the game to students, I recommend the following procedure:

• Show Alaska on a map of the United States, and explain that the Yup'ik Eskimos are one of the indigenous cultural groups.

• Explain the original purpose of kakaanaq in Yup'ik culture and how it helped develop hunting skills. Tell what materials the Yup'ik used in making the game and how it was traditionally played.

• Explain how the students will play kakaanaq in the classroom. Make a list of the good deeds or favors the winner will receive. Explain the scoring, and have the students practice scoring by completing the worksheet in figure 12.6.

After all the students have played the game, ask them what mathematics skills the Yup'ik people—and they—used in playing kakaanaq. Make a list of the mathematics used: measuring, adding, subtracting, strategizing, and problem solving.

Study each playing field carefully, and write down the correct interim scores for player A (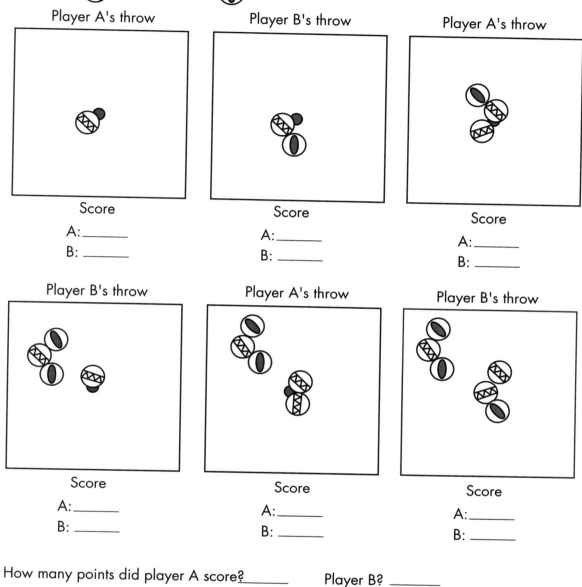) and player B ( ).

Player A's throw

Player B's throw

Player A's throw

Score

A: _____
B: _____

Score

A: _____
B: _____

Score

A: _____
B: _____

Player B's throw

Player A's throw

Player B's throw

Score

A: _____
B: _____

Score

A: _____
B: _____

Score

A: _____
B: _____

How many points did player A score?_____   Player B? _____

Fig. 12.6. A worksheet for practice in scoring kakaanaq

BIBLIOGRAPHY

Madenwald, Abbie Morgan. *Arctic Schoolteacher: Kulukak, Alaska 1931–33.* Norman, Okla.: University of Oklahoma Press, 1992.

Naske, Claus-M., and Herman E. Slotnick. *Alaska: A History of the 49th State,* 2nd ed. Norman, Okla.: University of Oklahoma Press, 1987.

Native American Games and Activities   13

Native Americans have been playing games of chance for ages and ages. About one hundred years ago, Stewart Culin, an anthropologist, traveled around the continent to learn about these games. He collected them in a book, *Games of the North American Indians* (Culin 1992), first published in 1907. He found that Native American children and adults played many games that were similar to one another. The players would toss six or eight objects—sticks, peach pits, or walnut shells, for example. The sticks or fruit pits were usually plain on one side and colored or decorated on the other. After each toss, the players counted the number these objects that fell with the decorated or colored side up and the number that fell with the plain side up. The players would earn points according to the way the objects fell. Their systems of scoring were often so complicated that the observer couldn't figure them out!

The game of "dish" and the "stick game" are simple forms of the Native Americans' games. The children learned the simple forms to prepare for the complicated games played by the adults.

Claudia Zaslavsky

The Game of "Dish"
Two or more players

Some of the Native Americans who played this game, often called the "bowl game," are the Seneca of New York, the Passamaquoddy of Maine, the Cherokee of Oklahoma, and the Yokut of California. The objects they used depended on what was available in their environment. The Seneca made buttons from the horns of elk. The Cherokee might have used beans. The Yokut used half-shells of nuts filled with clay or pitch. The players tossed the playing pieces on flat baskets that the women had woven.

Materials

- Four playing pieces (bottle caps, peach pits, walnut halves)
- Markers
- Fifty toothpicks or beans to keep score
- A wooden bowl, a pie pan, or a flat basket

Preparing the playing pieces

With the markers, decorate each playing piece on one side. Use a different pattern or color for each piece. You might want to decorate the pieces with Native American symbols (see fig. 13.1). The decorated side is called the *face* of the piece.

Playing the game

Place the toothpicks or beans in the center of the playing area. Put the playing pieces in the bowl. Decide in advance how many rounds to play and whether to impose a penalty if the pieces fall out of the bowl during play.

Fig. 13.1. Native American symbols

Take turns tossing the playing pieces into the bowl. Hold the bowl with both hands, and flip the pieces lightly in the air. Note whether they fall with the decorated sides (faces) up or down. Be careful not to let the pieces fall out of the bowl. If you have decided to do so, impose a penalty if that happens.

Scoring

- All four up   5 points
- Three up and one down   2 points
- Two up and two down   1 point
- One up and three down   2 points
- All four down   5 points

Count the number of points you earned. Take that number of toothpicks or beans from the pile, and place them next to you. The player with the greatest number of toothpicks or beans at the end of the agreed-on number of rounds is the winner.

Things to think about

Is the scoring system fair? Note that an outcome of four up is scored the same as four down. Is this a fair way to score the game? One way to find out is to toss one playing piece twenty times. Keep track of the number of times it lands up and the number of times it lands down. Are they about the same, or is one outcome more likely than the other? Repeat the experiment. Is the scoring fair or unfair?

Can you find six different ways that the pieces can fall with two faces up and two faces down? Label the pieces "A," "B," "C," and "D." Copy the chart in figure 13.2, and finish it.

A	B	C	D
up	up	down	down

Fig. 13.2. A chart for recording the ways four pieces can fall with two faces up and two down

Changing the Rules

Native Americans usually played these games with five, six, or more playing pieces. How would you score the game for five pieces? For six pieces?

The "Stick" Game

Two or more players

This Native American game is similar to the game of "dish."

Materials

- Four Popsicle sticks or tongue depressors
- Markers
- Fifty toothpicks or beans to keep score

Preparing the playing pieces

With markers, decorate each Popsicle stick on one side. Use a different decoration or color for each stick. You might want to decorate the sticks with Native American patterns or symbols (see fig. 13.3). The decorated side is called the *face* of the piece.

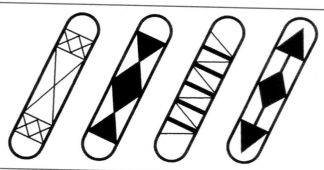

Fig. 13.3. Native American symbols and patterns

Playing the game

Place the toothpicks or beans in the center of the playing area. Decide in advance how many rounds to play. Take turns. Hold the sticks in one hand, and let them fall to the ground or the table.

Scoring

- All four up   5 points
- Three up and one down   2 points
- Two up and two down   1 point
- One up and three down   2 points
- All four down   5 points

Count the number of points you earned. Take that number of toothpicks or beans from the pile, and place them next to you. The player with the greatest number of toothpicks or beans at the end of the agreed-on number of rounds is the winner.

Things to think about

Is this a fair way to score the game? Try to think of a better way to score. Four sticks can fall in sixteen different ways that . Three ways are shown in the chart in figure 13.4. How many more can you find? Copy this chart, and finish it.

The Dream Catcher

The Ojibwa people, Native Americans of the Great Plains, have a traditional charm they call *bawa ji guun ahbee*, which translates to *dream catcher*. I learned about this tradition from a college mathematics teacher, who is the daughter of an Ojibwa elder. She also gave me the instructions for making the dream catcher. You will learn how to make your own dream catcher, a charm hung on a baby's cradleboard to bring good dreams to the sleeping child. Bad

#1	#2	#3	#4
up	up	up	up
down	up	up	up
up	down	up	up

Fig. 13.4. A chart for recording the sixteen different ways four sticks can fall

dreams were caught in the web, just as a spider's web catches and holds everything that touches it. Good dreams would slip through the web and make their way down the feather to the sleeping child. Following are the directions for making a dream catcher.

Materials

- A twelve-inch (30 cm) willow twig
- Five feet (1.5 m) of waxed string or dental floss
- One feather
- Four beads (one white, one red, one black, and one yellow)

If you cannot find all these materials, you may substitute others. Think about what would be suitable. You might use plain cord instead of waxed string. Instead of willow, use a thin twig from a different tree or bush. Soak it in water for a while to soften it before you bend it into a circle.

Making a dream catcher

1. Bend the twig into a circle, and wrap the ends with a piece of the string.
2. Tie four strands of string across the frame, evenly spaced. Each strand forms a diameter of the circle.
3. Tie all the strands together in the center (see fig. 13.5a).
4. Tie one end of a two-foot string to one of the cross strands. This is the weaving string. Start to weave, as shown in figure 13.5b.
5. As you weave, wrap the string tightly around each strand in a figure-eight knot. Follow the arrows and the numbers in the diagram in figure 13.5c.
6. Tie a feather onto a short piece of string. String the four beads onto the string, and tie the string to the center of the web, as shown in figure 13.5d.

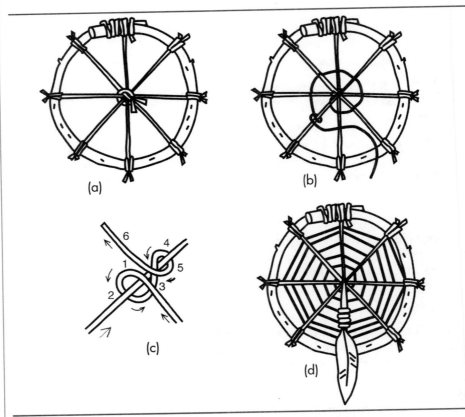

(a)

(b)

(c)

(d)

Fig. 13.5. Making a dream catcher

The dream catcher as a symbol

The dream catcher is round, like so many things in nature. Many people have thought that the circle is the most perfect shape. The circle is central in the lives of many Native American people. Here it represents the harmony of Mother Earth and nature. The string stands for the path of life. The beads are for the four directions. White is for the north, red is for the south, yellow is for the west, and black is for the east. The feather is for the effort needed to overcome the hardships of life.

Things to think about and do

- Why do you think the circle is so important in the culture and life of many Native Americans?
- Write a story to go with the dream catcher. Draw a dream catcher to illustrate your story. Read the story to a young child or to a friend. You may also want to read the book *Dreamcatcher*, by Audrey Osofsky (1992).
- Make a list, or draw pictures of things in nature that are round. Make another list, or draw pictures of round objects made by people. Why are they round and not some other shape?

REFERENCES

Culin, Stewart. *Games of the North American Indians.* 1907. Reprint, New York: Dover Publications, 1975; reprint, with an introduction to the Bison book edition by Dennis Tedlock, Lincoln, Neb.: University of Nebraska Press, 1992.

Osofsky, Audrey. *Dreamcatcher.* New York: Orchard Books, 1992.

The Mathematical Ecology of the Florida Seminole and Its Classroom Implications

14

James Barta

Anthropologists generally recognize that the indigenous people who first populated the areas now known as North and South America migrated over a land bridge connecting today's Siberia with Alaska. Although most researchers know that these first people carried weapons and tools, few consider that they, the ancestors of Native people today, also brought with them skills that involved the use of many mathematical concepts. Whether by counting (e.g., naming the quantity of game while hunting), measuring (e.g., creating clothing, weapons, and tools), locating (e.g., remembering sites for food collection and seasonal living), designing (e.g., decorating items of daily use), explaining (e.g., recognizing and describing cycles in nature), or playing (e.g., enjoying games of skill or chance), these ancient people relied on the application of mathematical principles to survive and succeed in their environment. Clearly, Native American children today are descendants of a people very capable of knowing and using mathematics!

This chapter describes traditional practices of one Indian nation, the Seminole, and opportunities that exist for teachers of Seminole children to provide effective and culturally relevant mathematics instruction. More globally, it is hoped that the reader will fully realize that all Native people traditionally used mathematics, that this knowledge must be shared with their students, and that with this awareness, every teacher can personalize her or his own mathematics curriculum to reflect the culture of the Native students whom they teach.

A BRIEF SEMINOLE HISTORY

When referring to the Seminole of Florida, one is actually acknowledging a people whose historical roots originate with the Paleo Indian hunters in the Southeast. Their history can first be identified within the Creek culture of what is now central Georgia and Alabama during the 1700s and 1800s (Sturtevant 1971). As time passed, the tribal composition expanded to include several ethnic subgroups of Creek people, as well as African Americans escaping slavery. Groups of Creeks formed a loose "confederacy" that linked various villages along the major creeks and rivers of the area.

The name *Seminole* is thought to have originated from the European mispronunciation of the Creek word *simanoli*, which meant *runaway* and seems to

The author wishes to acknowledge the Seminole Cultural Tribal Council for allowing him to pursue this study. He is particularly grateful to Pat Jagiel, Alice Snow, Mildred Bowers, June Tiger, Mary Johns, Martha Jones, Norman Tribbett, Martin Coyle, and Valerie Whiteside for their contributions and assistance. He also wishes to recognize John Morris Parker for his lasting support and many editorial suggestions.

have derived from the Spanish word *cimarron*, which has the same meaning (Garbarino 1988). Finding themselves displaced from their traditional lands, the Seminole were caught up in ongoing territorial disputes between Britain and Spain. Later, forces of the United States military were used against the Seminole in an attempt to secure the land that the government desired for political as well as economic purposes.

The Seminole are survivors. Three times during a thirty-year period reaching from 1812 through 1842, the United States waged war on them. Many Natives resisted with their lives the government's insistence that they give up their lands and accept forced relocation. The Trail of Tears is a legacy of those who were coerced to move to the Oklahoma Territory.

The harsh environment and the fact that U.S. soldiers were ill-trained to wage guerrilla-style warfare contributed to the government's disillusionment about achieving a successful completion of its military campaign. Eventually, with the Seminole population estimated to be fewer than five hundred, the soldiers were withdrawn. Those Natives remaining were left in peace to make their homes in the swamps and glades of central and southern Florida. The fact they were able to do so marks them as one of the few indigenous groups who for a time in history, after contact with outside interference, were able to maintain traditional ways of life, relatively free from outside intervention.

The Seminole were the first American Indian nation to be officially recognized by the United States government. Many Seminole pride themselves on the fact that they are the only Native group never to have signed an official "peace treaty" with the United States. Although traditional cultural practices are still evident, there is a growing concern, particularly among Seminole elders, that successive generations are slowly losing touch with the old ways—their ancestral culture. Today, a population of approximately two thousand Seminole live on several reservations across Florida.

MATHEMATICS AND CULTURE

In all cultures, evidence of mathematical ideas can be found (Ascher 1991), yet not all cultures know and use mathematics in similar ways. As cultures differ, so too do their individual ways of thinking and applying their mathematical understandings. Bishop (1991) has stated that many of the everyday activities of people, past and present, involve a substantial amount of mathematics. The universal tasks of counting, locating, designing, measuring, explaining, and playing were inseparably intertwined with other aspects of the Seminole culture and are the fundamental facets used in this article to probe traditional daily living practices of the Seminole.

Descriptions of traditional practices involving mathematics were given by a number of Seminole elders who were interviewed during a two-year research project conducted by the author in 1994 and 1995 and supplemented using related references and archival literature sources. Most interviews took place on the Brighton and Hollywood reservations. Research results have led to substantial insight into the traditional application of mathematical inventiveness within the Seminole culture. Through a study of these applications, as illustrated by the following examples, we can come to a better understanding of the Seminole culture and their early experiences using mathematics.

SEMINOLE
MATHEMATICS

Counting

Objects found in the natural environment were items whose quantity was described by counting. An examination of the numeric names suggests that Seminole counting is predicated on a base-ten system. No written symbols exist; rather, physical representations, such as seeds, pebbles, and knots, or finger gestures are used to describe quantities physically when necessary. When gestures come into play, however, counting appears to be done in groups of five; a person counts and touches each digit of one hand while counting from one to five. Counting on from six to ten involves retouching the digits already counted. Literal translations of the number names for six through nine seem to support the idea that at one time the numerical system had, at least in part, a subbase of five. The numbers 6 through 9 include a root word *apak*, meaning "to be with." Six, literally translated, means "with itself"; 7 means "2 with it" (i.e., with 5)"; 8 means "3 with it"; and so on. The number 11, or *palen hvmkontvlaken*, means one greater than 10. The numbers 12 through 19 contain a derivative of the word *kaken*, which means "to sit on." Hence, 12 literally means "2 sitting on 10," 13 means "3 sitting on 10," and so on. The words listed below are the counting names for the numbers 1–20 and 30–100 (J. Martin, personal communication, 23 January 1998).

1—hvmken	11—palen hvmkontvlaken
2—hokkolin	12—palen hokkolohkaken
3—tuccenen	13—palen tuccenohkaken
4—osten	14—palen ostohkaken
5—cahkepen	15—palen cahkepohkaken
6—epaken	16—palen epohkaken
7—kolvpaken	17—palen kolvpohkaken
8—cenvpaken	18—palen cenvpohkaken
9—ostvpaken	19—palen ostvpohkaken
10—paken	20—pale-hokkolen
30—pale-tuccenen	70—pale-kolvpaken
40—pale-osten	80—pale-cenvpaken
50—pale-cahkepen	90—pale-ostvpaken
60—pale-epaken	100—cokpe-hvmken

The Seminole referred to zero as "having nothing." Counting using extremely large numbers was not necessary in the daily lives of the people. Words for large numbers could be constructed in the language, however. For instance, the word for one million, *cukpe-rakko-vcule-hvmkem*, can be literally translated as "one old large hundred" (DeCeseare 1997). Although such quantities could be named, they do not appear to have been commonly used. To have too much of anything was considered culturally inappropriate. Doing so meant that one was greedy and must share with others in the camp. Reference was made to vast numbers, however, in other ways. For example, when asked to describe the number of stars in the sky, an elder stated, "There are so many of them that I could count for all of my life and I would die before I could finish."

Fractions were used by the Seminole to describe parts of whole objects. One elder described how after a deer hunt, her father would butcher the animal and give parts away to other families. The fraction that the other families received was determined by their need. This elder stated that words did exist for such fractions as one-half. She drew a circle with her hand on a table and then motioned as if cutting the circle in half. When asked whether the Seminole had a name for one-fourth, she explained that the word, when translated, literally meant one-half of one-half. Four was an important number to the Seminole because of the existence of four compass directions and four races of people (red, yellow, black, and white). An elder stated that when an accident occurs on the reservation, people become anxious because they believe that three more accidents may occur before they will be free from the danger. Martin (J. Martin, personal communication, 23 January 1998) stated that many activities are done four times, "apparently [to] signify completeness."

Traditional Seminole traded items for things that they needed, but they did not have any formal currency. They traded hides, pelts, and plumes of egrets at various trading posts that existed for the special purpose of buying with hard currency goods that were unavailable elsewhere and making these products available through traditional trades. Coins that had been obtained from British, Spanish, or American settlers were sometimes flattened, drilled, and sewn onto clothing as decorations. "Wealth" was determined by the number of domestic animals that one owned, and the women frequently possessed more than the men. Acquiring wealth was not a goal of the Seminole. Members of the tribe understood that food items and the like were to be shared. In fact, one was looked on in a negative light if one acquired too much. Excess acquisition indicated greed—that one was not properly sharing.

Locating

The Seminole located places within their environment by constructing mental maps. Landmarks that were familiar to most people acted as points of reference. Occasionally, they used maps drawn in the sand. The directions used to locate a position were given literal names. The cardinal direction east, for instance, was "the place where the sun comes up." When one slept, lying with one's head toward the east was important. To sleep in a different orientation, particularly with the head to the west, was considered unlucky, as corpses were aligned this way. The orientation of specific components of the green-corn dance, a sacred New Year's celebration, were set out again in relationship to the four directions. The fire pit was in the center of the dance circle, and from this point radiated camps for the clans attending, the game pole, the medicine pole, and so on.

Designing

Designs were abundantly evident in the Seminole culture. Ornate geometric linear patterns symbolizing some aspect of nature often adorned silver work, pottery, or finger-woven yarn sashes. The patchwork clothing worn by many women and a few of the men involved an implicit understanding of such transformational geometrical concepts as flips, slides, and rotations (see fig. 14.1). Tessellated strips of colored fabric that had been constructed with these motions were then stitched together as decorations to be applied to dresses and shirts.

Measuring

Seminole measurement involved the use of certain "standard units" found in the environment. Parts of the body provided convenient and suitable standard

Fig. 14.1. Illustration of a patchwork design created using a transformational geometric slide

units of measure. For instance, the distance from the nose to the end of an outstretched arm was used to measure units of cloth. The measurement used for constructing of the traditional Creek home, known as a *cuko* (see fig. 14.2), used the *po-cus-wv e-mv-pe* (pronounced ba-giz-u-ah e-mobi), which, when translated from the Creek language, means the "length of an axe handle."

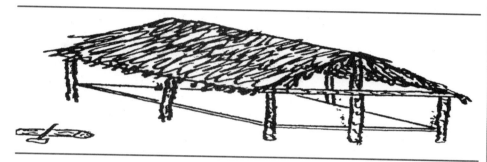

Fig. 14.2. Drawing of a Seminole cuko

Other "units of measurement," such as the number of paces between objects in the environment (distance) or the rate at which one traveled on foot or in a wagon to the trading post (speed), reflect how the Seminole people applied their mathematical intelligence to solve relevant everyday problems. Apparently, the Seminole possessed an uncanny ability to measure "by the eye." Exact units were not necessary, since minor fluctuations in linear tolerance created few real problems.

Time was often measured using the day or parts of the day as standard units. The green-corn dance, a festival typically celebrated during the second week in June to mark the beginning of a new year, was officially announced by runners to villages and isolated families two weeks before the date of the dance. Fourteen small sticks were hung, and one was taken down each day to mark the approach of the ceremony. When the last stick was removed, people knew that it was time to begin the ceremonial rituals (Cory 1896). When someone died and was buried, a fire was built at the foot of the tomblike structure in which the body lay. The enclosure was made of logs with a palmetto-leaf roof, and the fire was renewed at sunset for three or four days. At the end of the last day, the fire was allowed to burn itself out, and the deceased was allowed to rest in peace.

Telling time was related to the length cast by one's shadow, or perhaps by the shadow of a specific tree used as a reference point, at various periods of the day. For instance, if the shadow extended eastward, one knew that it was afternoon; the specific time depended on the length of shadow. Midday was known to be the time of day when no shadow appeared. Times for eating in the morning and the evening were fairly regular, although food was available throughout the day to anyone who was hungry.

Years were identifiable cycles consisting of two seasons, summer and winter. Each season was composed of six moons that were indicative of such seasonal environmental activities as animal or bird migration or the availability of black-berries or other foods.

Distances were expressed as a function of the time one took to travel from point *A* to point *B*. The speed or distance was determined by the mode of travel; walking was the slowest, travel in cattle-drawn wagons was faster, and canoe travel was the fastest. The distance from the village to the trading post may have been only a half-day's travel by canoe. Great distances were described as being so great that a man could walk his entire life and never arrive at the destination.

Explaining

Mathematics was also used to explain such concepts as age and the time between certain important events. One's age was calculated by counting how many times a particular season had occurred since one's birth. Therefore, one might be twenty-one "summers" old. This calculation amounted to counting how many two-season cycles, or years, had elapsed. One's age was also relative to those with whom one lived. A person might have lived many winters yet not be considered an "elder" if other family members who were older were still living.

Playing

Playing was an important aspect of the Seminole culture. During the sacred green-corn dance celebration, a type of stick ball was played. Teams of men and women or boys and girls attempted to strike the center pole with a leather ball stuffed with deer hair. If they were successful, they earned a point. The scoring was usually additive, with a point earned for each goal scored, but was occasionally subtractive, with points deducted from a predetermined opening score.

A game similar to the Appalachian mumble peg was played throughout the year. Children used a sharpened stick and took turns pitching it from various places on their body into a pile of sand. Points were scored when the stick landed vertically and remained upright in the sand. The knee bone of a cow was also used to play a tossing game. Each face of the bone earned players a specified number of points when the bone landed with that surface facing upward. Children added the combinations of numbers earned and played until a predetermined number was reached.

The activities just described represent only a few examples of mathematical applications in the daily lives of the traditional Seminole. Through such ethnomathematical analysis, we can glimpse how indigenous mathematics was an integral part of the Seminole culture and perhaps better understand how these people lived and conceptualized. With an awareness of the mathematical competence of these people comes an opportunity for teachers to help today's Seminole youth take pride in their cultural heritage. The six aspects of mathematical activity used to demonstrate the mathematical resourcefulness of the Seminole can be used by all teachers to formulate mathematical activities appropriate for the culture of their own students. Several of the activities described have been successfully applied at Ahfachkee Seminole School located on the Big Cypress Reservation near Clewiston, Florida.

CLASSROOM IMPLICATIONS AND ACTIVITIES

The activities that follow include short summaries of the ethnomathematical information acquired through previous research. The mathematical concepts presented are first explained through examples that illustrate the concepts' importance to traditional Seminole culture. Once students become aware of these cultural implications, the teacher can use their understanding as a bridge to demonstrate how the concepts are used today. Although specific objectives are presented, the activities are designed to encourage the students to begin asking their own questions about the mathematics being taught and seeking their own solutions through approaches that reflect their current and historical cultural identities. These activities have been created to supplement existing commercial curricula that seldom, if ever, illustrate the long history of mathematical application and prowess among the Seminole.

Counting Activity

Objective

Students will analyze patterns in the Seminole counting system and compare and contrast it with systems from other cultures, including the Hindu-Arabic system.

Summary

The Seminole have a counting system that is based on groupings of ten. When Seminole people have been observed counting from one to ten on their fingers, though, some count from one to five by pointing to the fingers on one hand. Next, as they count from six to ten, they again count the same fingers on the same hand, a practice that indicates that groupings of five were also important. An analysis of the written numbers 6–10 illustrates the custom of placing a derivation of the counting numbers (1–4) in front of the part of the number indicating a group of five to compose the sum. Originally, the Seminole did not have a written language but did communicate numbers verbally. These words are the counting names for the numbers 1–14 and 20–300. Traditionally, the quantity 0 was described as "having nothing." "Infinity" and vast numbers were described as being numbers so large that you could count all your life and never reach them.

These words are the counting names for the numbers 1–20 and for the decade names for 10–100:

1—hvmken	11—palen hvmkontvlaken
2—hokkolin	12—palen hokkolohkaken
3—tuccenen	13—palen tuccenohkaken
4—osten	14—palen ostohkaken
5—cahkepen	15—palen cahkepohkaken
6—epaken	16—palen epohkaken
7—kolvpaken	17—palen kolvpohkaken
8—cenvpaken	18—palen cenvpohkaken
9—ostvpaken	19—palen ostvpohkaken
10—paken	20—pale-hokkolen

30—pale-tuccenen	70—pale-kolvpaken
40—pale-osten	80—pale-cenvpaken
50—pale-cahkepen	90—pale-ostvpaken
60—pale-epaken	100—cokpe-hvmken

Activity

Study the number names, and describe any patterns that you see. How does this system of your ancestors compare with the Hindu-Arabic system that you have learned in school? Write the words for the numbers that you use in daily life—for instance, your address, age, telephone number, and the like.

Measuring Activity

Objective

The students will create a number of units of measure from objects found in their environment and will use them to measure objects around them.

Summary

The Creek word *po-cus-wv e-mv-pe* (pronounced ba-giz-u-ah e-mobi), or "length of an axe handle," was a standard building unit in house construction. Parts of the body (e.g., length between the nose and one extended arm) were also frequently used when measuring smaller things. Many objects found in nature were used as "standard" units.

Activity

Forage outdoors, and select sticks that you think would be a suitable length for an axe handle. Using a nonstandard unit of measurement mutually agreed on as a class—for instance, a hand length or a foot length—measure your handles. Compare the findings, and discuss whether one length appears to be the most appropriate. Can you test this conclusion by measuring and finding the numerical average? Using available natural materials and exploring the concept of proportionality, can you construct a model of a traditional Seminole home, called a *cuko*?

Locating Activity

Objective

The students will create maps of locations important to them using self-selected frames of reference.

Summary

The Seminole located places in their surroundings using a variety of techniques. Although written maps were apparently not created, occasionally maps were drawn in the sand. Many people made mental maps and used familiar landmarks as points of reference. The directions used to locate a position had literal names. For instance, the cardinal direction west was named "the place where the sun goes down."

Practices in daily life were often influenced by one's position in relation to one of the four cardinal directions. When one slept, it was important to lie with one's head toward the east. To sleep in a different orientation—particularly with the head to the west, since corpses were aligned in this direction—was considered unlucky. Whether planning a village camp or constructing a site for the yearly ceremony of the green-corn dance, the Seminole used mapping concepts that incorporated a number of mathematical ideas.

Activity

Interview your tribal elders to determine how a Seminole camp of several families would have looked. Draw a map of the camp, and include its important features. How large will your map be? What scale will you use to represent the selected features? If you had to explain to someone the location of a particular point or place on your map, how would you do it?

Designing Activity

Objective

The students will explore geometric transformations, such as flips, slides, and rotations, as they design Seminole patchwork patterns.

Summary

The Seminole often decorated their silver work, pottery, or finger-woven yarn sashes with lively geometric patterns. The traditional dress of the Seminole was often decorated with strips of colored fabric constructed to create beautiful representations of such aspects of nature as rain, fire, lightning and thunder. These patterns were designed by flipping, sliding, or rotating patches of colored cloth and sewing them together for decorations on skirts or shirts. This application of geometrical concepts, referred to as *transformation geometry*, gave the Seminole a distinctive style of clothing design that is still in evidence today.

Activity

Using a variety of colored strips of construction paper ranging in size from one to several centimeters in width, students can paste them horizontally to card stock in alternating patterns. The students then cut vertical strips; reorganize them by making flipping, sliding, or rotating motions; and reglue them to an additional piece of card stock. The variety of possibilities is endless, and students may want to decorate their own notebooks or other items of daily use with these cultural and mathematical patterns.

Explaining Activity

Objective

Students will create ways to describe the passage of time (e.g., the customary minutes, hours, days, or years) and compare these with techniques that the Seminole used to explain similar aspects of time.

Summary

The Seminole used mathematics to explain such concepts as age and the time between certain important events. The year was described as a cycle of two seasons, and one's age was calculated by counting how many of these cycles had occurred since an event happened. Rather than be fifteen "years" old, one might be fifteen "summers" old. Time was often told by observing the shadow cast by a tree that stood in some prominent position where all could see it. The direction of the shadow told the people whether it was before noon or afternoon. Many other aspects of time were explained by using patterns in nature as "clocks."

Activity

Students are asked to consider how they could describe the passage of time (e.g., the customary minutes, hours, days, or years) if clocks and watches did not exist. The students are given time to create and construct their own timepieces cooperatively using common materials found in the classroom. They are then presented with information about traditional ways in which the Seminole kept track of time. Finally, they are asked to describe the similarities and differences among the methods that they have invented and those used by the Seminole.

Playing Activity

Objective

Students will apply mathematical operations to keep score while playing a Seminole children's game. Later they will be encouraged to invent their own games, rules for playing, and scoring procedures.

Summary

Playing games was an important aspect of the Seminole culture. The games ranged from running and throwing to those of chance. Score was kept by adding or subtracting points. Two games of chance involved tossing a sharpened stick or a cube-shaped bone and awarding points for the way that it landed. For one game, children used a sharpened stick and took turns pitching it from various places on their body into a pile of sand. Points were scored when the stick landed vertically and remained upright in the sand. Another game used a bovine knee bone shaped like a cube. The children named each side of the cube with a number, such as 1, 2, 5, or 10. Some of the six faces were represented by the same number, and children earned the points specified on the side of the bone facing upward after a toss. The children added the combinations of numbers earned and played until a predetermined number was reached.

Activity

Students in teams are given blocks of wood and are asked to write values on the faces. They should carefully consider the probability of their block's landing with a particular face up and should name the faces with appropriate values. They should take turns tossing the block individually but should accumulate points collectively as they play to a predetermined sum. When the total is reached, all the players win! Children may also enjoy opportunities to invent their own system of playing and scoring using new blocks. They should be asked to describe the mathematics that they have used in their newly invented games.

REFERENCES

Ascher, Marcia. Ethnomathematics: *A Multicultural View of Mathematical Ideas.* Belmont, Calif.: Brooks/Cole Publishing Co., 1991.

Bishop, Alan. *Mathematical Enculturation: A Cultural Perspective on Mathematics Education.* Norwell, Mass.: Kluwer Academic Publishers, 1991.

Cory, Charles B. "Hunting and Fishing in Florida." In *A Seminole Sourcebook,* edited by William C. Sturtevant, pp. 7–40. New York: Garland Publishing, 1896. Reprint, 1971.

DeCeseare, Richard. "An Investigation of Muskogee Creek Indian Counting Words." *International Study Group on Ethnomathematics Newsletter* 13, no. 1 (1997): 2–3.

Garbarino, Merwyn S. *The Seminole.* Indians of North America series. New York: Chelsea House Publishers, 1988.

Sturtevant, William C. "Creek into Seminole." In *A Seminole Sourcebook,* edited by William C. Sturtevant, pp. 92–128. New York: Garland Publishing, 1896. Reprint, 1971.

Walking in Beauty

Three Native American Elementary Preservice Teachers Rediscover Mathematics

15

James Barta

> My people (Kiowa/Diné) were never discussed by the teachers and certainly not in math class, even though most of my classmates were also Native American. I grew up thinking that there must be something wrong with me and my people; I thought my people didn't know anything about mathematics and seldom ever used it.
>
> —Shirley Reeder

This chapter describes early mathematical experiences of three Native American preservice teachers and explains the ways in which Native American culture can be used to connect with mathematical instruction.

Shirley is a Native American student studying at a university in the western mountain region of the United States. She and two other Native American students—Ann, who is Shoshone, and Candace, who is Diné—are preservice education majors who someday hope to return to their reservations and teach. Their stories concerning the mathematical educations that they have received illustrate that the instruction was seldom connected with their cultural heritage. These students, for whom in their first languages the word *mathematics* did not even exist, struggled to see the meaning and relevance of the instruction they received. The difficulties that they faced as mathematics students may not have been as dependent on their misunderstanding of the subject as on the ineffective and inequitable teaching techniques used by their teachers.

Shirley, Ann, and Candace have been given a second chance to learn the mathematical concepts that they did not acquire as young mathematics students—concepts that they will need to know if they are to be successful teachers. They are getting this chance through the way that they are now being taught to teach mathematics, specifically, through connections with culture. They are discovering that many of their routine, culturally important activities have required them to use mathematics. As a result, they are now beginning to realize that they have "always been mathematicians." For them, mathematics has come alive because they have learned to look at its relationship with culture and in doing so have become more confident in their own abilities to understand it, use it, and teach it.

The NCTM stresses through its *Standards* documents (NCTM 1989, 2000) the importance of connecting mathematics instruction with a context that is

RELATING MATHEMATICS WITH CULTURE

The author is indebted to Ann, Candace, and Shirley, who over the course of two years have shared their stories and experiences. May they always walk in beauty!

familiar to the learner. Even though many educators are aware of this precept, teachers do not always clearly understand the connection of culture as it relates to their mathematical instruction. Educators unfamiliar with ethnomathematics, or relating mathematics with culture, may benefit from an understanding of it as they expand their mathematical perceptions and more effectively instruct their students.

Ethnomathematics is a term used to express the relationship that exists between mathematics and culture. D'Ambrosio (1987), who has been credited with coining the expression, explains it as "the mathematics which is practiced among identifiable cultural groups, such as national-tribal societies, labor groups, children of a certain age bracket, professional classes, and so on" (p. 45). For the three Native American preservice teachers featured here and for many other students, mathematics was taught to them with little or no cultural connection. In classes where teachers did describe mathematics used by people outside the classroom, the representatives often portrayed distant cultures, frequently having European histories.

Traditionally in mathematics classrooms, the relevance of culture has been absent from daily content and instruction. In many classrooms, students learn rote procedures either that have no cultural context or that reflect situations more familiar to mainstream populations. Barta (1995), in referring to mathematical instruction void of cultural connections, made this statement (p. 3):

> Inaccurate instruction misleads all children about the richness of mathematical history and to a degree about the people who have populated this planet. For minority children, who as a group have not realized the same level of mathematical success as European American students in our classrooms and who are often underrepresented in higher level math courses and professions requiring significant mathematical competence (Reyes & Stanic, 1988), the effect of this misinformation may be particularly devastating. Many of these children simply do not realize that they are mathematically capable and that they do in fact possess a long and rich mathematical heritage.

Lave (1988) challenges such practices and insists that the culture of the student and the way that he or she thinks are inseparable. Bishop (1991, p. xi) explains mathematics as a "way of knowing," and Wilder (1978, p. 26) further explains it as follows:

> Those people who use mathematics—the "mathematicians"—are not only the possessors of the cultural element known as mathematics but, when taken as a group in their own right, so to speak, can be considered as the bearers of culture, in this case mathematics.

Thinking and learning occur in a social context among people; the child later internalizes this context in his or her individual development (Vygotsky 1981). The way that a child thinks is reflective of the "social form" in which those thoughts occurred. If one realizes that all students bring to class a cultural identity through which they see and understand, then the idea of connecting mathematics and culture becomes synonymous with effective and equitable education. Basing instructional dialogue on a context familiar to the students not only enables them to see how the concept or idea relates to their daily life but also supplies an extension through which they can expand their understanding. They become able to see the relevance of the concept in contexts that previously were beyond their cultural repertoire.

Much of human learning prior to the development of formal, mass education resulted from a combination of direct experience and interaction with others who had greater competence in the task being completed or the problem being solved. This intellectual scaffolding (Rogoff and Gardner 1984) furnished by the "expert" allowed the learner to hear an explanation of the task, see a demonstration of its application, and participate with guidance in carrying it out until they, too, were fully competent. All this learning occurred in a context in which both the "teacher" and the "student" had familiarity.

Native American students sometimes postpone participation in an unfamiliar task until they feel sufficiently competent from watching someone who is capable of successfully completing it. Additionally, to be effective, the instruction needs to be relevant and have personal meaning. Gregory Cajete, a Tewa Indian and educator, explained the process of traditional instruction among indigenous people as follows (Cajete 1994, p. 33):

> The living place, the learner's extended family, the clan, the tribe provided the context and source for teaching. In this way, every situation provided potential opportunity for learning, and basic education was not separated from the natural, social, or spiritual aspects of everyday life. Living and learning were fully integrated.

Native American students today have much to teach us if we listen carefully to them and if we are willing to adapt our instructional practices to align better with their ways of knowing and learning. When we show our students that we value them because we value their culture, we become more sensitive and effective educators. We enhance avenues of communication as both we and our students benefit from the ensuing cultural transfer of ideas and perspectives. As we learn to recognize students' culturally specific uses of mathematics, they learn that we, too, as cultural representatives, have developed unique ways of understanding and applying mathematics.

STORIES OF DISCOVERY

The experiences of three Native American preservice teachers will be described to illustrate how these educators are discovering themselves mathematically through an examination of their cultures.

Ann—Hoop Dancing

Ann is a member is of the Shoshone-Bannock and Isleta Pueblo Nations. She is known as a gifted dancer and often competes in powwows across the region, as well as performs at celebrations and special community events. She does "hoop dancing," which Ann describes as follows (Abeyta 1997):

> Hoop dancing is done with a number of circular (10, 15, 21) hoops that traditionally were made of willow or cane but can be constructed today of plastic. Originally dancing through the hoops brought blessings as the dancer passed through each hoop opening.

Using a number of the hoops while keeping count to the song that is sung and the cadence of the drum beat, Ann whirls and arranges the hoops to represent a number of animals or objects. The "pictures" that she creates with the hoops honor those animals or objects that they are made to represent.

The hoops that Ann has constructed are proportional to her body, and she explains this customization to her curious students. She encourages her students to think about how this application of measurement is similar to the stories that they hear about how such English standard units as the "foot" were determined. Ann's explanation included the following (Abeyta 1997):

I measure the size (diameter) of the hoop so that it is the same distance as from my foot to my knee. This way I know that as I dance I can easily slip through them. My people often measured (length) using parts of their bodies. People everywhere measure and often choose their own units from their environment.

In her teaching, Ann is beginning to incorporate more of her Native American culture, and this approach has been well received by both students and teachers. For instance, as she dances, she may ask her second-grade students to count carefully the number of hoops that she is using. The students add or subtract and call out their sum or difference as Ann completes each figure.

Ann also incorporates Native American symbolism when she works with common fractions. She does so by explaining the "medicine wheel." The medicine wheel is an important, colorful symbol to many Native American people. It is a circle divided into quarters, each of which is a particular color. Ann states that the wheel is often used to represent Mother Earth, or the world, and that the four colors represent the colors of the black, red, white, and yellow people who inhabit it. Each color also represents a particular cardinal direction and a related spiritual quality or supernatural host. Although the medicine wheel is recognize by many Native Americans, not all Indian nations may use or interpret the colors similarly.

Using an overhead projector, Ann questions her students about the amount of the area of the wheel that is covered with a color or combinations of colors. The students eagerly describe their responses. Much of Ann's instruction involves encouraging students to solve culturally specific "problems" cooperatively. She might ask them to determine the shape of a house that they would build if they were limited to a certain total perimeter. She may phrase the problem in this way (adapted from Zaslavsky [1989, p. 20]):

> Imagine you belong to a group of people who believe that when you reach 15 years of age you must build your own home. The home must only be one level and because you will collect all that you will need to build your home (sticks, branches, grasses) from where you live, a rule of your people states that the distance around the outside of your home cannot exceed 24 units. In what shape will you build your home so that you have inside the greatest amount of living space?

Children who are given a 24-unit (feet, yard, or meter) rope loop can actively create various shapes and stand inside them. Some students will be quite surprised to discover that if the perimeter's length is held constant, the shape of their house will make quite a difference in its area! Using grid paper, students can draw their shapes and invent ways of determining the related areas. Later, a discussion can be held to share what has been discovered. The circle is usually the shape that students determine will provide the greatest amount of living space, or area. Ann may talk now about the traditional homes of the Shoshone people, the tepee and the wicciup, both structures with a circular base. Did Ann's people build those traditional homes in that shape as a fluke, or were they quite mathematical?

Candace—Navajo Rug Weaving

Candace is a member of the Navajo Nation, whose people describe themselves as *Diné*, a term that means "the people." The traditional craft of the Diné is rug weaving. Candace's teachers are her elders, who pass their knowledge on to her as together they create a rug. Like many other Native American students, Candace is proud of her heritage and feels honored and privileged to speak with, and learn from, her many teachers on the reservation. From them, she has learned much about traditional uses of mathematics that she hopes to explain to her students.

Candace (John 1997) has learned that most rugs have proportionate dimensions; typically they are made so that their width is one-half their length (e.g., 3 ft. × 6 ft., 4 ft. × 8 ft.). Although numerous blanket or rug designs exist, symmetry is important and is nearly always present. The rugs can be folded in half to make doubles, or reflections, and even be folded in half again to make quarters. The patterns that the weavers design are often not preplanned but evolve as the work progresses. The weaver can sense when her or his work is in harmony and the artist is said to "walk in beauty."

Since much traditional teaching occurs through stories or songs sung by the *ha taali*, or singer–medicine man, much of what Candace has learned has been conveyed in this way. Candace has learned that the Diné describe the quantity zero as *a di'n*, meaning "nothing." Infinity, or *doo nii t'eeh da*, is translated as "it doesn't end." The Diné' sometimes refer to themselves as the "five-fingered people." When counting, they describe the quanty five as *ashdla*, whose root derivative, *ala*, means "five fingers—the hand." The Diné counting words from one to ten are (1) *ala'aii*, (2) *naaki*, (3) *taa*, (4) *dii*, (5) *ashdla'*, (6) *hastaan*, (7) *tsostsid*, (8) *tseebii*, (9) *na'hasttei*, and (10) *neesnaa*.

A multiplicative process is used to count further by tens, as Candace describes in the following:

If you want to say twenty, you first take the first syllable of two which is Naa and you take the word "diin" which means "in the tens" ... the (Navajo) word for twenty or "Naa diin" is saying two times ten. One hundred would be "Neezna diin" meaning ten times ten." Words to describe very large numbers, such as millions, were not known to the Navajo until contact was made with the Spaniards and traders. We really didn't have a use for it (millions), but we did have the word "t oo a hoyoi" which means "many or a lot."

In addition to counting, Candace states that for the Diné, certain numbers have special significance.

The numbers four, nine, and twelve are such examples and all have special meaning. The number "four" typically is one number important to many Native Americans. For the Navajo, there exist four sacred mountains, four worlds, colors, female guardians, plants, directions, winds, elements, original clans, seasons, etc. Being a matrilineal society, the number nine is important. Women go through nine months of pregnancy. The hogan and the sweat house is considered to be the womb of Mother Earth and in there is where we live or go to pray to her and learn how to walk the ways of the Diné. As we live in the womb of Mother Earth, there are nine songs in the Hogan song which tell of our creation. The number twelve is also significant because there are 12 moons, which mark the repeating pattern of the four seasons. In one of our sacred prayers called the "Earth Up Bringing Prayers," we pray to our universal family who are twelve beings (Mother Earth, Father Sky, Female Mountain, Folding Darkness, Early Dawn, Talking Spirit, Calling Spirit, White Corn, Yellow Corn, Corn Pollen, and Corn Beetle.

The language of the Diné has is no formal word for *mathematics*. Rather, mathematics is what Candace's people do. Imagine the difficulty that may arise if teachers of traditional Navajo students are not aware of this cultural difference! Candace concludes that she has learned a lot about who she is culturally through her ongoing study of her culture's connection with mathematics. She enjoys this type of learning and looks forward to the day when she can help mathematics come alive for her students.

Shirley—Native American Beadwork

"Topping the list of my least favorite subjects was mathematics," remarks Shirley when asked to reflect on her educational experiences as a child

(Reeder 1997). She believes that much of her problem resulted from the fact that her teachers did not help her see the relevance of what they were teaching. As a future teacher, she does not want her students to suffer the same fate. As a mother of five, she is already working hard to show her children that many of their Native American crafts and traditions embody mathematical principles.

Shirley is Kiowa. Her people reside in western Oklahoma, where she hopes to return to teach after graduation. Shirley is a gifted beader who uses her beading tools and supplies to quickly create beautiful beadwork sashes and rosettes to adorn clothing items that she and others wear.

In the summer, when she is not in school, she and her family travel to, and participate in, powwows that are held on weekends across the region. She does much of the beadwork for the dance costumes for herself and her family. She states that when she sits down to bead, she seldom holds in her mind an image of the final product. Rather, she begins to bead and follows her intuition, which guides the development of the design. She knows when the pattern that she is making is "not right," because she will have trouble, perhaps with the needle and thread. She understands that this difficulty may be a sign that the particular design should not be continued, so she will stop it and start anew.

According to Shirley, she never really knew that she used so many mathematical concepts in the practice of her traditional crafts until she began to learn about ethnomathematics. She now states that not only does she realize that she understands and uses many mathematical applications that formerly were incomprehensible but also that her interest in finding out more about her native culture has been rekindled. Shirley explains as follows:

> When you do loomwork, it is important to know that you always must string your loom so that you have an odd number of beads in each (horizontal) row. This is because the median (bead) acts as the center point. You use it to flip your pattern so that one side will mirror the other. We can use this idea with children learning math. How many times as a student did you hear your teachers tell you that what you do to one side of the equation, you must do to the other? This beadwork example makes this concept real for children. I tell my own children to think of that middle bead as an "equal sign" so that their patterns are symmetrical.

Competence in counting and computing are skills that Shirley believes are vital when doing beadwork. She says that first the various colors of beads being used must be continuously counted. As certain geometrical patterns begin to unfold, Shirley realizes that she will have to add or subtract beads to create the preceding rows. She explains that beadwork is often done by recognizing numerical patterns. The order and pattern of the Fibonacci sequence can be seen in many beadwork designs. Shirley continues her description:

> When I am showing my children how to create a border say for their mocassins they are making, I have them skip count. Even the young children catch on quickly as they count by threes, fours, or fives depending on the width of border they are making. My older children have learned to think of the multiplication facts. They see the arrays made by the beads and can quickly mentally calculate the number of beads they have used.

Opportunities abound for children who study mathematics using beadwork. Besides applying the operations already described, they can explore measurement (perimeter and area) and can describe quantities of beads (colors and composition) using common fractions, decimal fractions, ratios, and percents. They can explore geometrical, or transformational, applications using flips, slides, and rotations to create unique patterns. Technology will now allow your student to test design patterns on the computer using programs typically used for doing needlepoint. Bradley (1992) has worked with Native American students learning Logo to "draw" their design on the computer; later they work with tribal elders who teach them traditional ways to do the beading or weaving. The students learn the expected mathematical concepts through cultural applications. Additionally, for the first time, many of them have also begun to learn how to practice the arts and craft of their people. The elders are pleased with this cultural preservation; and as their elders share their knowledge, experience, and wisdom, the children gain both culturally and mathematically.

TO WALK IN BEAUTY: CONCLUSION

A *hataali*, or medicine man, speaking to Candace, told her this (John 1997):

> The true principle of life comes when you stop looking for it and start living the life the Saah Naghai Beki Hoozhon (your holistic being) intended for you. Your holistic being must be in harmony from the tip of your toes to the top of your head—the traditional way of life is to walk in beauty and in balance.

The Diné believe that to live a good life, one must be aligned with who and what one is culturally. The harmony that results when one understands one's cultural relationship with the world allows one to "walk in beauty." Connecting mathematics and culture is important for all students. For the teacher of Native American students, it presents an opportunity to use the cultural characteristics and traditions of the child to explain the personal relevance and meaning of mathematics.

Educators communicate thoughts and ideas through their cultural perspectives. The challenge for the successful teacher is to learn to understand better how his or her culture has influenced the way that he or she understands and uses mathematics. Empowered with this awareness, such a teacher may realize the necessity of using students' cultural perspectives to communicate mathematical thoughts and ideas. When teachers help their students become aware of the many ways that their ancestors have used and still continue to use mathematics, they endow those students with the realization that as direct descendants, they, too, possess mathematical ability.

REFERENCES

Abeyta, Ann. Personal Communication, 23 June 1997,

Barta, James. "Reconnecting Maths and Culture in the Classroom: Ethnomathematics." *Mathematics in Schools* 24 (2) (March 1995): 12–13.

Bishop, Alan. *Mathematical Enculturation: A Cultural Perspective on Mathematics Education.* Norwell, Mass.: Kluwer Academic Publishers, 1991.

Bradley, Claudette. "Teaching Mathematics with Technology: The Four Directions Indian Beadwork Design with Logo." *Arithmetic Teacher* 39 (May 1992): 46–49.

Cajete, Gregory. *Look to the Mountain: An Ecology of Indigenous Education.* Durango, Colo.: Kivaki Press, 1994.

D'Ambrosio, Ubiratan. "Reflections on Ethnomathematics." *International Study Group on Ethnomathematics Newsletter* 3 (September 1987).

John, Candace. Personal Communication, 25 June 1997.

Lave, Jean. *Cognition in Practice: Mind, Mathematics, and Culture.* Cambridge, U. K.: Cambridge University Press, 1988.

National Council of Teachers of Mathematics (NCTM). *Curriculum and Evaluation Standards for School Mathematics.* Reston, Va.: NCTM, 1989.

———. *Principles and Standards for School Mathematics.* Reston, Va.: NCTM, 2000.

Reeder, Shirley. Personal Communication, 25 June 1997.

Reyes, Laurie, and George Stanic. "Race, Sex, Socioeconomic Status, and Mathematics." *Journal for Research in Mathematics Education* 19 (January 1988): 26–43.

Rogoff, Barbara, and W. Gardner. "Adult Guidance of Cognitive Development." In *Everyday Cognition: Its Development in Social Context*, edited by Barbara Rogoff and Jean Lave, pp. 95–116. Cambridge, Mass.: Harvard University Press, 1984.

Vygotsky, Lev S. "The Genesis of Higher Mental Functions." In *The Concept of Activity in Soviet Psychology*, edited by J. Wertsch. Armonk, N.Y.: M. E. Sharpe, 1981.

Wilder, Raymond L. *Evolution of Mathematical Concepts*. London: Open University Press, 1978.

Zaslavsky, Claudia. "People Who Live in Round Houses." *Arithmetic Teacher* 37 (September 1989): 18–21.

Finding Cultural Connections for Teachers and Students

The Mathematics of Navajo Rugs

16

Recent research (National Research Council [NRC] 1991; Rosser 1995) has documented the shortage of scientific professionals, both in the general population and more dramatically among ethnic and cultural minorities. Minorities are still underrepresented in advanced mathematics and science courses at the secondary, undergraduate, and graduate levels. Too often, ethnically and culturally diverse students find school mathematics irrelevant to their lives. The inclusion in the elementary and middle school curriculum of multicultural mathematics can motivate diverse students to participate in mathematics and thus have opportunities to obtain the prerequisite skills necessary for success in advanced mathematics courses.

All students benefit from learning about multicultural mathematics. Multicultural curricular materials add interest to mathematics, provide opportunities for interdisciplinary connections, give context to mathematics lessons, and offer students the chance to appreciate the universal nature of mathematics. When students learn that virtually all cultures use mathematics in their daily lives, they begin to broaden their view of what mathematics is. Teaching students about different cultures' contributions to the development of mathematics can help debunk the myth that mathematics is of European origin and a male domain.

Elementary and middle school teachers have the responsibility for including multicultural mathematics early on. However, preservice and in-service teachers often are not adequately trained to develop and teach multicultural lessons. Although such issues as race bias or gender inequities are frequently addressed with preservice and in-service teachers, the hope is that they come away from these experiences treating all their students with equal respect and concern. However, these experiences are not usually content specific and hence do not transfer easily into their own teaching, except in an affective sense.

To meet the needs of teachers and teacher educators who want to integrate culture-based activities into their teaching, I have designed three related mathematics lessons that teach significant mathematical content while paying tribute to Navajo culture. To support the integration of mathematics with other subject areas, I have included information on the history of Navajo weaving, appropriate children's literature, and connections with science and with other cultures. These lessons were designed for use by all students. The activities serve as models of lessons that teachers might design for their own classes. Although the lessons described here focus on one culture's use of mathematics, in my classes I use examples from several cultures at appropriate places in the curriculum.

Joan Cohen Jones

HOW AND WHEN TO INCLUDE A MULTICULTURAL CURRICULUM

Decisions about how and when we integrate a multicultural curriculum are crucial to its success. Frankenstein (1997, p. 16) contends that cultural examples "should *not* be presented as a kind of 'folkloristic' five-minute introduction to the 'real' mathematics lesson; rather these hidden contributions should be presented in their material context...." Moreover, multicultural activities should be integrated into the regular curriculum rather than used as "add-on" enrichment lessons. For students to respect and appreciate the mathematics contributions of other cultures, multicultural activities should be connected with what they are presently learning in mathematics and other disciplines. Finally, the pedagogy used with a multicultural curriculum is an important component of its success. A multicultural curriculum works best in student-centered, active classrooms where students work in cooperative learning groups (Sleeter 1997).

When I teach multicultural mathematics activities to my teacher education classes, I follow a three-step process. First, to model the integration of multicultural activities into the curriculum, I connect the activities with the unit that we are studying. Second, I ask my students to put on their "teacher hats," assume a different role, and analyze how and why I included a multicultural unit at this point in the curriculum. Third, I ask the students to discuss how they would integrate the same activities at the level at which they intend to teach.

For the Navajo rug activities described here, I usually introduce the topic by reading children's literature about Navajo weaving. Through discussion, we connect weaving with the history of the Navajo people and discuss its place in their culture. To make the activities relevant to my students' level of knowledge, I include the activities as part of a gender-equity unit, to encourage discussion about what should be considered mathematical knowledge (Frankenstein 1997) and women's "hidden" contributions to mathematics. We discuss how traditional women's work, such as weaving and sewing, requires a great deal of mathematical knowledge and how this fact contrasts with the myth that women are not good at mathematics. The activities are thus placed in context for my students at a level that they find meaningful.

The preservice teachers then reflect on how they can integrate the Navajo rug activities at the elementary or middle school level. We usually decide to present them as part of an interdisciplinary unit, perhaps on the American West, the development of the railroad, or the history and displacement of Native Americans during the nineteenth century.

RESULTS FROM FIELD TESTING

These lessons have been successfully field-tested with preservice teachers in mathematics education courses, with practicing teachers during in-service programs, and with a fourth-grade class in a small midwestern city. Activities 1, 2, and 3 were presented to my elementary-school-level introductory mathematics education and mathematics methods classes. The preservice teachers enjoyed the lessons, particularly activity 1. Linking fractions with something as tangible as Navajo rugs actually enhanced their understanding of fraction concepts. One preservice teacher stated that she had never really understood fractions before doing this activity.

After completing activities 1, 2, and 3, I asked my students to reflect in their journals on their personal beliefs or ideas about teaching and learning that had changed as a result of the Navajo activities. The results were overwhelmingly positive. One first-year student commented on integrating mathematics with other subject areas:

> I thought of math as very separate from history and [E]nglish, but now I know that math adapts to other subjects. I now believe it is very important to incorporate math into more than just school work. We need to relate it to games, architecture, art, and measurement.

Another student wrote about the importance of including culture in the classroom:

> With all the many cultures being brought into our nation and schools, I also feel we need to include them in the classroom. I feel it is a great advantage to experience and gain knowledge of the many types of people and their heritage. Everyone knows how curious children are. By expanding their focus on all the many cultures, it could be a bonding experience for many of the children.

The activities were presented to twenty in-service teachers of grades K–5. Most of the teachers thought that their students would benefit from the activities, modified to their own level. They particularly appreciated the connections with children's literature and U.S. history that the activities foster. A few teachers questioned the appropriateness of the topic for their geographic area, since no Navajo people live in the area in which the teachers work and live, although other Native American tribes live nearby. I responded to the teachers' questions by emphasizing that (*a*) these lessons were to be considered as models for what the teachers themselves could create and (*b*) students benefit from learning about the mathematics of many different cultures. When presented with several cultural examples, students can look for cross-cultural mathematical concepts and ideas among them. For example, the use of symmetry and geometric shapes in textile design is quite common, found in Native American weaving and beadwork, West African weaving and resist-dyed fabric, and Hmong needlework.

Several teachers expressed concern about the possibility of offending minority students with multicultural lessons. One teacher told of her experience as a first-grade teacher on a Navajo reservation. One day, she took her class for a nature walk and gathered plants, rocks, and a horned toad for the class terrarium. The next day, most of the children in her class were absent. After consultation with the school principal and a Navajo schoolbus driver, the teacher realized that her actions had unwittingly offended the Navajo children and their families. The Navajo believe that their ancient people return to earth in different forms. The horned toad might have been a relative come back to earth. The schoolbus driver, who was also a spiritual leader, performed rituals to cleanse the teacher and the classroom so that the children could return to school. The teacher's method of finding out what went wrong is a helpful model for preservice and in-service teachers. She consulted community members for accurate information and accepted the practices and beliefs of community members as well as their solution to the problem. Although I have been unable to verify independently the teacher's interpretation of Navajo beliefs, I believe that this example reveals the key to successfully including multiculturalism in the classroom and school. A multicultural curriculum works best when it is interactive. Parents, grandparents, and community members are valuable resources and should be consulted and included whenever possible.

The Navajo weaving lessons were also integrated into the mathematics curriculum of a fourth-grade class, one hour a week for about four weeks. Culture plays an active role in this class. About one-fourth of the students are Hmong. Their teacher includes Hmong culture, art, history, and literature in the curriculum, as well as information about many other cultures. The students were already familiar with Navajo culture and were very excited to learn more about the Navajo, especially in relation to mathematics. They particularly enjoyed designing their own Navajo rugs. Because the class had not yet studied fractions, activity 1 was modified accordingly. The students had already had some exposure to symmetry and polygonal shapes and revisited these concepts in activities 2 and 3.

A BRIEF HISTORY OF NAVAJO WEAVING

The Navajo learned to weave around A.D. 1600 from their Pueblo Indian neighbors. At first, the Navajo wove clothing and blankets in browns, beiges, and grays, the natural colors of wool. Their weaving quickly progressed. They began to use vegetable dyes and to weave blankets of great intricacy and beauty. Some of these blankets, known as *chief blankets*, were traded to the chiefs of other Native American tribes.

Navajo weaving changed dramatically in the mid-1800s. In 1863, the Navajo were hunted down and arrested by the United States Army. By 1865, they were imprisoned at Fort Sumner (Bonvillain 1995). During their imprisonment, their weaving was influenced by Anglo traders. These traders, particularly John Lorenzo Hubbell (Taylor and Taylor 1993), suggested that the Navajo weave rugs instead of blankets, include a border around each rug, and use commercial dyes to create brighter colors. Anglo traders and railway officials continued to influence Navajo weaving until the early part of the twentieth century. By this time, the quality of Navajo weaving had deteriorated. In recent years, Navajo weavers have returned to traditional patterns and vegetable dyes, producing rugs that are collected all over the world.

IDEAS FOR MATHEMATICS LESSONS

Navajo rugs have rich mathematical connections. They often use polygonal shapes to create designs that contain symmetry, tessellations, and transformations (see fig. 16.1). Most teachers do not have easy access to Navajo rug designs, but several books provide Navajo rug patterns that can be adapted for classroom use. *Navajo Rugs and Blankets: A Coloring Book*, by Chuck Mobley and Andrea Mobley (1994), presents historical facts about Navajo rugs and depicts simplified rug designs. *North American Indian Designs for Artists and Craftspeople*, by Eva Wilson (1984), contains intricate designs as well as historically accurate information about each design. Bulletin-board trimmers using Navajo designs are available commercially. I have used these trims for in-service workshops by cutting them into small pieces and distributing them so that each participant has a Navajo design. The following are some ideas for mathematics lessons based on Navajo rugs.

Fig. 16.1. Example of a Navajo rug design

From *Southwest Indian Designs*, by Madeleine Orban-Szontagh (New York: Dover Publications, 1992); used with permission

Activity 1: Navajo Rugs and Fractions, Decimals, and Percents

NCTM's *Curriculum and Evaluation Standards for School Mathematics* Addenda Series, Grades K–6: *Fourth-Grade Book* (Burton et al. 1992) contains a quilt-pattern activity that can be adapted to Navajo rugs. Give each participant a sheet of ten-square-by-ten-square grid paper and samples of Navajo patterns. Ask participants, "How many squares are on the grid? How can we represent one square on the grid by a fraction, a decimal, and a percent? How can we represent the entire grid by a fraction, a decimal, and a percent?" Ask each participant to create his or her own Navajo rug design on the grid paper and then color the design with three different colors. Next estimate, then calculate, what fraction, decimal, and percent of the rug is covered by each of the three colors.

Ask participants why weavers might need this information. One reason is to help the weaver estimate how much yarn of each color is needed. For example, if the weaver knows how many skeins of yarn are needed to weave the entire rug and that one-third of the rug will be red, then the weaver will dye one-third of the yarn red. Weavers usually spin their own yarn and dye it before they weave the rug. If the weaver does not dye enough yarn red and needs to dye an additional batch, the second batch may vary in color from the original because it comes from a different dye-lot.

A discussion of the colors used in the rug naturally extends to a discussion of the concepts of perimeter and area. Ask each participant to estimate the perimeter and area of his or her rug and the area of each of the colors used in the rug, then to compare the methods and results with those of other participants. What conclusions can be drawn? Develop standard techniques for calculating perimeter and area, using unit squares.

Finally, encourage the participants to compare their rug designs. Have them group their rugs according to different characteristics and to discuss the groupings. To conclude the activity, make a "blanket" of all the rug designs for a bulletin board.

Activity 2: Navajo Rugs and Symmetry

This activity is a good follow-up to activity 1. Most Navajo rugs contain line symmetry (Yei rugs do not). Some also contain rotational symmetry. Using the Navajo rug designs created during activity 1, ask the participants to make a vertical fold in their designs or hold a ruler vertically so that the grid-paper "rug" is divided into two equal parts. Ask whether the designs are exactly the same on each half of the rug, that is, have vertical line symmetry. Make a horizontal fold in the rug to check for horizontal line symmetry. Ask the participants to rotate their rugs to determine whether the design stays the same through one, two, three, or four rotations. Through class discussion, encourage the participants to generate their own definitions for line and rotational symmetry.

Activity 3: Polygonal Shapes in Navajo Rugs

This activity is designed for younger children but has also been used with preservice and in-service elementary-grades teachers. The goal is to identify the different shapes found in Navajo rugs and get a sense of angle size. Provide several examples of Navajo designs. After the participants have identified the polygonal shapes in the designs, give each individual a small square and right triangle cut from cardboard. Discuss the similarities and differences between the square and right triangle, emphasizing both square corners and other corners. Ask the participants to analyze the angles or corners in the rug samples. Are they bigger, smaller, or the same as the square corners of the cardboard square? Next, ask the participants to use the cardboard square and right triangle to find other squares, right triangles, and corners—square and otherwise—in the rug samples, in the classroom, and at home.

CONNECTIONS WITH CHILDRENS' LITERATURE AND SCIENCE

Several wonderful children's books have been written about the Navajo, both fiction and nonfiction. Suggested books include *The Goat in the Rug*, by Charles L. Blood and Martin Link (1990), *The Navajo*, by Nancy Bonvillain (1995), *The Magic of Spider Woman*, by Lois Duncan (1996), *The Navajos: A First Americans Book*, by Virginia Driving Hawk Sneve (1993), and *Songs from the Loom: A Navajo Girl Learns to Weave*, by Monty Roessel (1995). These books can be incorporated into an integrated unit that includes mathematics, literature, geography, social studies, and science.

Science activities can center on the sources of Navajo native dyes. The participants can conduct hands-on investigations to dye cloth, yarn, or paper. Some easily obtainable dye sources include rose hips, brown onion skin, red onion skin, blue flower lupine, and sunflower petals.

CONNECTIONS WITH OTHER CULTURES

The story of Navajo weaving is similar to the experiences of other cultures. For example, Hmong needlework changed dramatically after the Hmong people were interned in refugee camps in Thailand. Both Navajo weavers and Hmong embroiderers memorize patterns, never writing them down. Navajo weavers leave a *spirit line*, a break in the border of the rug, so the spirit of the weaver is not trapped inside. Hmong embroiderers sew a portion of the border in a different color of thread for similar reasons.

Similarities also exist between Navajo rugs and Amish quilts. Both Navajo rugs and Amish quilts have intricate geometric patterns that have line and rotational symmetry. Both were initially created from available raw materials, wool and scraps of cotton. And both Navajo rugs and Amish quilts were used for warmth rather than decoration. One important difference is that the Navajo rug is woven by the female members of one family, whereas the Amish quilt is sewn by female family members along with other community members.

CONCLUSION

The NCTM's *Standards* documents (1989, 2000) present a vision of mathematics that is accessible to all students. Yet culturally and ethnically diverse students usually perform more poorly in mathematics than their peers in the mainstream culture. As educators, we must develop strategies that reach all students. Multicultural mathematics activities that use culture to make connections with typical mathematics topics can motivate culturally and ethnically diverse students to investigate and gain respect for their own cultural heritage while learning significant mathematics content. Multicultural mathematics enriches the study of mathematics for all students by illustrating the global nature of the development of mathematics, as well as the mathematics concepts and ideas that are shared across cultures. The activities discussed here provide a model for teachers and teacher educators who are striving to include culture and cultural context in their own classes.

REFERENCES

Blood, Charles L., and Martin Link. *The Goat in the Rug.* New York: Macmillan, 1990.

Bonvillain, Nancy. *The Navajo.* Brookfield, Conn.: The Millbrook Press, 1995.

Burton, Grace, Douglas Clements, Terrence Coburn, John Del Grande, John Firkins, Jeane Joyner, Miriam A. Leiva, Mary M. Lindquist, and Lorna Morrow. *Curriculum and Evaluation Standards for School Mathematics* Addenda Series, Grades K–6. *Fourth Grade Book.* Reston, Va.: National Council of Teachers of Mathematics, 1992.

Driving Hawk Sneve, Virginia. *The Navajos: A First Americans Book.* New York: Holiday House, 1993.

Duncan, Lois. *The Magic of Spider Woman.* New York: Scholastic, 1996.

Frankenstein, Marilyn. "In Addition to the Mathematics: Including Equity Issues in the Curriculum." In *Multicultural and Gender Equity in the Mathematics Classroom: The Gift of Diversity,* 1997 Yearbook of the National Council of Teachers of Mathematics (NCTM), edited by Janet Trentacosta and Margaret J. Kenney, pp. 10–22. Reston, Va.: NCTM, 1997.

Mobley, Chuck, and Andrea Mobley. *Navajo Rugs and Blankets: A Coloring Book.* Tucson, Ariz.: Treasure Chest Publications, 1994.

National Council of Teachers of Mathematics (NCTM). *Curriculum and Evaluation Standards for School Mathematics.* Reston, Va.: NCTM, 1989.

———. *Principles and Standards for School Mathematics.* Reston, Va.: NCTM, 2000.

National Research Council. *Moving beyond Myths: Revitalizing Undergraduate Mathematics.* Washington, D.C.: National Academy Press, 1991.

Roessel, Monty. *Songs from the Loom: A Navajo Girl Learns to Weave.* Minneapolis: Lerner Publications Co., 1995.

Rosser, Sue Vilhauer. *Teaching the Majority: Breaking the Gender Barrier in Science, Mathematics, and Engineering.* New York: Teachers College Press, 1995.

Sleeter, Christine E. "Mathematics, Multicultural Education, and Professional Development." *Journal for Research in Mathematics Education* 28 (December 1997): 680–96.

Taylor, Cobin, and Betty Taylor. "Native North America." In *Textiles: 5000 Years,* edited by J. Harris, pp. 264–72. London: Harry N. Abrams, 1993.

Wilson, Eva. *North American Indian Designs for Artists and Craftspeople.* New York: Dover Publications, 1984.

Native American Patterns

Cathy A. Barkley

The day-to-day activities of people, both past and present, involve many mathematical applications. Most of the applications have not been considered as mathematics because they do not involve lengthy calculations or complicated formulas. Bishop (1991), as quoted in Barta (1996), has identified six universal activities as mathematical practices by any culture: counting, measuring, designing, locating, explaining, and game playing.

Designing, or organizing shapes, patterns, and colors on surfaces is one way that people have used one of these basic mathematical practices. The discipline of mathematics often includes the study of patterns. Patterns can be found everywhere in nature, and often these patterns are copied and adapted by humans to enhance their world further. Nowhere is this practice more evident than in the study of designs by indigenous peoples. The study of mathematical ideas of native or indigenous peoples is referred to as *ethomathematics*. One particular aspect of mathematics that is evident in all cultures is the use of geometric designs.

Designs and patterns on utilitarian objects have been found in all cultures throughout history. Such patterning or decoration does not make an object any more useful. A basket, for example, is no more or less useful for carrying things whether it is plain and utilitarian or whether it is woven with an intricately patterned design. However, all cultures have invested time and energy to create aesthetically pleasing designs. Many examples of these patterns are found on the pottery, clothing, and containers of Native Americans and in their living spaces. The Yup'ik, or Eskimo, people of Alaska have traditional names for their geometric parka-trim patterns. To describe their designs, they use words that mean mountains, windows, sled runners, and "put little things together."

STUDENT PATTERNING ACTIVITIES

Introduce the idea that the structure or balance of a design is described by mathematicians as a set of rigid motions, or *isometries*. For various figures, different motions will move the figures about the plane but leave the figures unchanged except for their orientations. Display a copy of the Onondaga wampum belt shown in figure 17.1, which shows a pattern that uses repetition. The belt's design is an example of one of the simplest patterns, a *translation*. Another example of a translation, or slide, is shown in figure 17.2, a Ute beadwork design on a belt. The figure remains the same as it slides along a straight line. To demonstrate a translation, have students draw a figure and then slide it along a straight path.

Discuss the fact that to decorate an object, Native Americans used translations to create a strip pattern called a *frieze*. One popular Native American frieze pattern found in beadwork designs is based on the isosceles triangle. Have individual students create their own frieze by drawing a basic pattern, such as with the isosceles triangle, and translating it as many times as possible on a copy of the grid in figure 17.3.

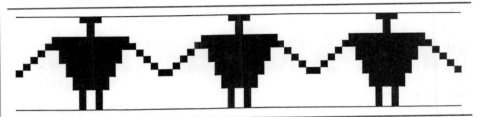

Fig. 17.1. Repetition in an Onondaga wampum belt

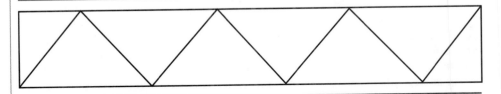

Fig. 17.2. A Ute beadwork design on a belt

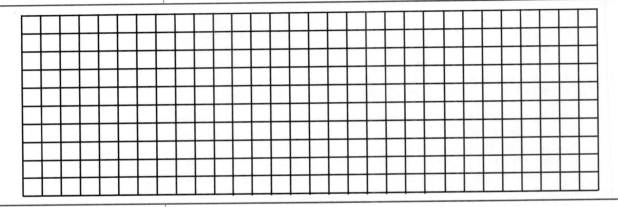

Fig. 17.3. Grid for students' use in creating their own patterns

At this point, introduce some of the vocabulary of geometric tessellations. The structure and balance of a design is described as an *isometry*, or rigid motion. Different motions on a figure produce different patterns. These motions can be categorized when examined more closely. A translation, examined previously, is one of these categories. Another type of motion is called a *reflection*, which, like a mirror, produces the mirror image of a figure. Reflections can be vertical, horizontal, or both. Display to the class the example of a repetitive horizontal reflection shown in figure 17.4 and that of a repetitive vertical reflection that appears in figure 17.5. When the reflection is repeated in both the horizontal and vertical directions, a pattern appears, as students can see in the example in figure 17.6.

Have students create another frieze by drawing a pattern and repetitively reflecting it along a horizontal or vertical line. Distribute a blank copy of the grid in figure 17.3 to each student for this purpose. You might have the students display their creations around the classroom or in a class book. Students can also create patterns by using two motions, that is, translating the image and then reflecting it. This motion is known as a *glide reflection*, an example of which is shown in figure 17. 7. Have each student draw a pattern on a copy of the blank grid (fig. 17.3) and then use the glide reflection to create a new frieze. In a whole-class discussion, encourage students to compare and contrast their creations.

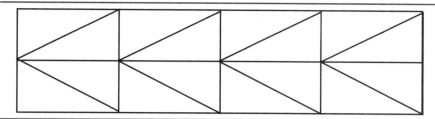

Fig. 17.4. A repetitive horizontal reflection

Fig. 17.5. A repetitive vertical reflection

Fig. 17.6. A reflection that is repeated in both a horizontal and a vertical direction

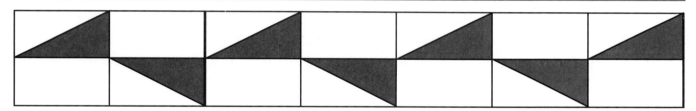

Fig. 17.7. A glide reflection

A *rotation* turns a figure about a point. Figure 17.8 shows an example of a half-turn, or 180-degree rotation. Figure 17.9 repetitively translates figure 17.8 to create a pattern.

Have students create another frieze by drawing a new figure on a grid copy (fig 17.3), rotating it 180 degrees, then repetitively translating the new drawing along the horizontal grid. To introduce a writing component, ask students to write a short paragraph identifying whether translations, reflections, or rotations were used to create the patterns in figure 17.10.

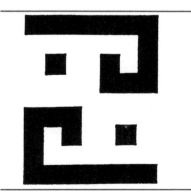

Fig. 17.8. A half-turn, or 180-degree rotation

Fig. 17.9. Repetitively translating figure 17.8

Fig. 17.10. Figures for students to analyze and describe their creation

CONCLUSION

Creating their own geometric patterns modeled on Native American designs gives tactile learners a chance to succeed in mathematics class and enlivens the related mathematics lessons for *all* students. The use of these activities brings multiple rewards for students: heightened interest and involvement in mathematics; an awareness and appreciation of the mathematical contributions of many cultures throughout history; and a sense that they, too, are able to accomplish meaningful mathematics.

Editors' note: A four-page reproducible student activity titled "Native American Patterns" and authored by Cathy Barkley is found in *Student Math Notes* (National Council of Teachers of Mathematics, Reston, Va., September 1997).

BIBLIOGRAPHY

Ascher, Marcia. *Ethomathematics: A Multicultural View of Mathematical Ideas.* Pacific Grove, Calif.: Brooks/Cole Publishing Co., 1991.

Barta, Jim. "Mathematical Thought and Application in Traditional Seminole Culture." *International Study Group on Ethomathematics Newsletter* 11 (June 1996).

Billstein, Rick, Schlomo Libeskind, and Johnny W. Lott. *A Problem-Solving Approach to Mathematics for Elementary School Teachers.* 5th ed. New York: Addison-Wesley Publishing Co., 1993.

Hargittai, Istvan, and Magdolna Hargittai. *Symmetry: A Unifying Concept.* Salinas, Calif.: Shelter Publications, 1994.

Naylor, Maria. *Authentic Indian Designs.* New York: Dover Publications, 1975.

Stanley-Miller, Pamela. *Authentic American Indian Beadwork and How to Do It.* New York: Dover Publications, 1985.

Washburn, D. K., and D. W. Crowe. *Symmetries of Culture—Theory and Practice of Plane Pattern Analysis.* Seattle, Wash.: University of Washington Press, 1988.

Weyl, Herman. *Symmetry.* Princeton, N.J.: Princeton University Press, 1952.

Zaslavsky, Claudia. *The Multicultural Math Classroom: Bringing In the World.* Portsmouth, N. H.: Heinemann, 1996.

Teaching Mathematics Skills with String Figures

Gelvin Stevenson
James R. Murphy

18

Weaving string figures are a visually pleasing and wonderfully tactile way of learning to appreciate complex consequential phenomena. The visual, manual, and aesthetic dimensions of making string figures attract students who are otherwise bored with standard mathematics classes. Among its many virtues, work with string figures increases concentration and attention span. It empowers and gives confidence to students, as well as teaches them to persevere. Weaving figures is relaxing and has been reported to decrease violent behavior among students, who often pick up their strings when they need to relax. The weaving process teaches students that shortcuts and quick fixes do not work. Conceptually, work with strings shows how to proceed step by step to get somewhere. Accuracy is paramount: one false step, and the figure just does not appear. Abstraction becomes relevant and natural, especially when the instructions are codified. And weaving strings emphasizes the importance of accuracy and order. Operationally, the process teaches about addition, order, transitivity, laws of substitution, and other mathematical operations.

For more than a decade, coauthor James (Jim) Murphy has pioneered the practice of using string figures to teach mathematical skills at LaGuardia High School of Music and the Arts in New York City (Bluestone 1986). He has subsequently developed and copyrighted a methodology, Learning Hands, to maximize the learning value of string figures. The process expands several classic figures into systems that contain numerous variations of the basic weaves, plus some additional weaves. Students learn to alter these figures in several systematic ways, then to analyze the patterns that they create. Once they have mastered the basic weaves, they are presented with a written system of describing, or coding, these figures and their variations. Students can either adopt that codification system or develop their own. A more complete discussion of this aspect of the learning of string figures occurs in the "Codification" discussion below.

The Learning Hands family of techniques structures the experience of learning string figures to optimize the learning that takes place. In other words, it shapes, channels, organizes, and fortifies the learning that takes place naturally. It makes the ad hoc learning that occurs with all string figures explicit and purposeful without ignoring the pure fun of it all.

Murphy has taught these techniques in some of his mathematics classes, workshops for teachers and students, after-school programs, and summer camps. He has shown that the process works for students anywhere along the ability spectrum. In addition, he has taught the system to a blind student and to exercise therapists who use it with clients who have arthritic hands. Parts of the

This chapter is reprinted with permission from articles appearing in the *Bulletin of the International String Figure Association* (Stevenson 1995; Murphy 1997). For more information about the International String Figure Association, write to P.O. Box 5134, Pasadena, CA, 91117, or visit the association's Web site at www.isfa.org/~webweavers.

system have been incorporated into the Indigenous Science and Math curriculum in the schools at Akwesasne Cultural Center on Saint Regis Mohawk Indian Reservation in Hogansburg, New York.

This chapter opens with a description of what students learn from working with string figures in general and from the Learning Hands method in particular (Stevenson 1995). It concludes with a condensed presentation of Murphy's (1997) first lesson, in which students learn to create increasingly complex designs by systematically altering the movements used in forming a traditional Native American string figure, Osage Two Diamonds. The experience, fanned by its creator's enthusiasm, has ignited sparks of learning in countless students whose flames of intellectual inquiry had been doused by the rough-and-tumble life of New York City. Murphy, who like coauthor Gelvin Stevenson, is Cherokee, taught both his students and his colleague that string figures are a journey without end, a journey that spans the globe and all of human history.

SOME EXPLORATORY THOUGHTS

We begin with some intriguing quotes and anecdotes:

Music and mathematics both create their own worlds of abstraction, which they then proceed to explore. (Walton 1995)

Strings represent the world of pure reasoning—like Euclidean Geometry. They require no prior knowledge. You don't need to master a lot of information or a storehouse of knowledge to begin. You just use your brains and your hands.... You don't even have to know how to read. (Hapgood 1995)

Learning from start to finish a string figure requires complex mental operations that if mediated as a process to the youth may help them apply these processes to other similar situations. (Emerson 1988, p. 1)

The natives state that string figures were used to memorize popular chants and to recall tales. I was told that children were taught to make them and were obliged to memorize the accompanying chants before they were instructed in the sacred lore of the tribe and the art of carving tablets. (Métraux 1940, p. 354)

The system of writing was not explained until the students' memories had been perfectly trained. (Métraux 1957, p. 186)

The figures offer an excellent means for developing manual dexterity and coordination between brain and hand, and present a challenge to personal inventiveness too, since they are capable of infinite variations. (Jayne 1906, quoted in Moore [1988, p. 18])

String figures are fundamentally human and inviting, unthreatening and approachable. Conceptually, string figures are abstract and concrete at the same time, and they contain infinite possibilities. As such, string figures can draw practically anyone into amazingly complex ways of doing and thinking. This allure makes them extremely effective in helping students learn mathematics and other complex subjects or simply improve their thinking. Although mathematics can intimidate people from the outset, string figures invite them into a world where learning is intrinsic, natural, immediately rewarding, and fun. Any intimidation from string figures lasts only until one puts the string on one's hand or until one masters the first figure. The "hook" of string figures is just part of their magic. In developing this system, we have amplified and systematized those qualities to optimize the learning that occurs.

The following sections discuss three types of outcomes from learning and performing string figures—(1) psychological and emotional, (2) conceptual, and (3) operational.

Psychological and Emotional Outcomes

Calmness

The importance of psychological and emotional outcomes became apparent to coauthor Gelvin Stevenson one day during a visit to one of Murphy's string-figure classes at LaGuardia High School. When a graduate of a previous class dropped in, Stevenson—always trying to understand how string figures affect people—asked him, "What did you get out of string figures?" He answered quickly and easily, "It made me less violent." This response stunned Stevenson and lead to a great deal of thought and many conversations with people. The idea seemed plausible, although not all the mechanisms and nuances of the outcome were immediately evident. How did the calming effect occur? What about string figures had the capacity to make people less violent?

Without exception, the people that Stevenson interviewed validated the possibility of string figures' having a soothing effect. But the most striking confirmation came from a friend, who said that manipulating string figures reminded him of a lot of the principles that he had learned in anger-management therapy. These lessons included the following:

- In working with string figures, one has no other target for one's anger. Everything is between the weaver and the string. What one gets is completely and utterly one's own responsibility. One creates one's own outcomes.

- In making string figures, no quick solutions or quick fixes are available. One must go through several steps to reach a desired end. There are no short cuts.

- Working with string figures releases tension and promotes relaxation. One must concentrate on manipulating one's hands. Other concerns melt away and slide out of one's consciousness.

- Weaving string figures is meditative. Some people call creating string figures a *moving meditation*—it has the ability to focus one's attention on what one is doing with one's hands. In this way, the process can be likened to playing a musical instrument or repeating a mantra or prayer.

As striking as this set of findings may seem, they are only a few among the positive psychological and emotional effects that string figures can have. Here are nine others.

Confidence

When some people first encounter string figures, they think that they could never do them, even those who are skeptical at first can soon be weaving interesting and intricate figures. Almost invariably, the successful completion of an unfamiliar figure is accompanied by a big smile and wide eyes, and occasionally even by a loud exclamation and an impromptu dance. A student once explained how the process works for her: "You're not afraid to make a mistake. Just do it, and if it's wrong, you correct it, then you start again." Another student said, "It helped me with my Auto Skills (a computer mathematics program). After strings, I wasn't afraid to go in and try it. I knew I could figure it out." The confidence spawned by working with string figures can spread to other areas of a person's life.

Self-Evaluation

One knows whether one has gotten a figure right. One does not need a teacher or any other external authority figure to validate one's effort.

Patience

One cannot hurry a string figure. No short cuts can be taken, and the speed at which one can manipulate the strings is limited. Steady, thoughtful, step-by-step process, in contrast, yields rewarding results. Ultimately, string figures must be unraveled, and the more impatient one is when unraveling a figure, the longer the undoing takes. It cannot be hurried.

Attention Span

A restless student can put down a book or story after a few sentences and pick it up again. The contemplative activity has little to encourage the student to push ahead, except possibly enjoyment. But one cannot stop creating a string figure in the middle and then pick it up again. One wants to complete a figure in process, and that desire can often stretch one's attention span (Storer 1994).

Concentration

One's mind can wander while reading, but not when doing string figures. If one fails to concentrate on the string figure, it will not come out right. In discussing string figures and concentration, Emerson explains, "I can slow down my thinking in order to (input) information in an orderly and/or step-by-step fashion. I realize that time plays an important role in thinking. In fact, if I let time pass while I am engaged in thinking processes, the product of thought is always better in terms of quality and quantity. Knowing this I willingly restrain thinking that is too fast or impulsive" (1988, p. 3).

Consistency

One must be consistent—must do the same thing every time—to arrive at the same figure.

Attention to Detail

One must be exact in doing a string figure. Precision and accuracy of thinking are absolute requirements. Coming close does not count; neither does, in modern youth-culture parlance, "whatever."

Awareness of Spatial Relationships

String figures teach or enhance an awareness of spatial relationships. "[W]here a thing exists in relation to something else becomes critically important in weaving and enjoying figures" (Storer 1994).

Thoughtfulness

Work with string figures often makes students more thoughtful because it prompts them to reflect on what they are doing, what they have done, and why things turn out the way they do. Viewed another way, students are engaging in "restraint of impulsive thinking." (Emerson 1988, p. 4).

Conceptual Outcomes

Manipulating string figures enhances several conceptual abilities, a few of which are elaborated in this section.

Giving and Following Intricate Instructions

Teaching string figures requires exact and precise instructions, especially if the teacher must describe the method rather than show it, as can be verified by anyone who has written such instructions or tried to tell someone over the telephone how to create a figure. Even when teacher and learner are together physically, the instructions should be as precise as possible. This intrinsic need for exactness creates demands on both the teacher and the learner, forging a reciprocal relationship between the two. "The burden of the proof and exactness is on the giver so that the receiver may reciprocate. On the other hand, [the receiver of an instruction has to] know how to interpret instructions so that the correct result is achieved. In order for [a learner] to follow instructions they must 'slow down' their thoughts in order to see and understand the processes involved in the string games" (Emerson 1988, p. 4).

Properly handled, the need for precision can be a useful way to teach people to communicate with exactitude. Novice string-figure teachers quickly learn the ineffectiveness of imprecise instructions, such as, "Then you put that over there, and, using this finger, pick up that one, and then you do this, and then this." Having students teach students encourages exactness of expression. As they learn to me that learning to give instructions more precisely, students are helped to think more clearly.

Abstraction

String figures have a real but abstract quality. They are concrete enough to be easy to see and understand yet are still abstract. This abstract quality may help students become comfortable with other abstract areas, such as mathematics, computer language, or music. Learning to recognize and manipulate the abstractions inherent in string figures is greatly enhanced by codifying the figures—learning or developing various symbols or codes and using them to record instructions for string figures (see the section on codification). Codifying string figures propels students ahead in their understanding of abstractions and abstract languages, including computer languages. This gain is especially pronounced when they create their own symbolic way of writing string-figure instructions.

In Learning Hands, weaving string figures is envisioned as the manipulations of a complex, abstract reality. By working with families of string figures and systems of weaves, students learn to perform systematic manipulations of this complex, abstract reality.

Order of Operations

The strict sequencing of operations—detailed order—is important in mathematics and practically every other part of life. Producing each string figure requires an exact order of operation. Weaves 1, 2, 3 work for a given figure, but weaves 1, 3, 2 do not. AC is not CA. "Seeing and doing things in a sequential manner is vital to understanding many things. At times a loose and over generalized process of sequencing is not enough. . . . Certain things won't work unless strict sequencing of operations are made" (Emerson 1988, p. 3). One student articulated this idea with simple elegance: "You can't jump into the middle or end without doing the beginning."

Rule Making

To weave string figures, one must follow rules, which in turn imply order and relationships. Only the simplest weave is disembodied. One creates interesting figures by performing weaves in particular relationship to one another. "All relationships are governed by some kind of rule; knowledge of rules and rule making can greatly improve one's sense of relationships or behavior" (Emerson 1988, p. 4).

Systemic Thinking

Weavers think in terms of systems. For instance, in Diamonds, Ten Men, Nauru Island, and other such string-figure families, one can often substitute components of one system into another. Tom Storer, a professor of mathematics at the University of Michigan and a pivotal figure in the art and mathematics of string figures, refers to this capability as "respect for systematology" (1994).

Mixed-Modality Linkups

Several steps are embodied in creating every string figure. Much of the learning involved in Learning Hands—as well as the creativity of this art—comes from identifying or isolating those steps and then substituting weaves or moves between different figures or between different systems of figures. This activity is a good way to learn modular thinking, and it is analogous to certain principles of computer programming, in which sections of code or language are extracted from one program and used in another.

Codification

Codifying one's string-figure activity is an important part of the Learning Hands process. The experience of learning or developing the steps and then using a method of notation to remember and communicate them takes string figures to a whole new level of learning. It helps students see the structures and systems inherent in the figures. Once they have done so, they can more easily manipulate the figures and develop new figures and new systems. They can manipulate the abstract reality of string figures on a higher level of abstraction. By *codification*, we mean using words, letters, symbols, drawings, or any other representation to indicate particular actions or movements of the fingers and hands. Codifying can be done in many ways, ranging from Tom Storer's path-breaking system of highly detailed codes to Murphy's more-symbolic system.

The process is simple. First, students learn to produce the figures by emulating their teacher. Second, they learn or figure out how to write the instructions for creating that figure. Finally, they learn to read the codes—to go directly from the written or spoken codes to doing the weaves. In this way, they learn the relationship between abstract formulas and the figures that result on their hands. One can think about the codification of string figures as a way of allowing people to create their own mathematics—a kind of do-it-yourself mathematics.

A very simple form of codes that Stevenson developed at the Fortune Society helped his students when they took computer classes. They reported that they were no longer afraid of the abstract symbols that they saw on the monitors. They had a sense that those symbols represented something real. "Strings are just like computer language," said one of the Fortune students. "Once you understand the code, you're talking computer language. You don't sweat, get nervous or confused, when you see that computer language."

Operational Outcomes

Manipulating string figures can also teach many specific operational mathematics concepts. The essence of a set of instructions is a formula. Our methodology uses matrices to show how outcomes vary with different combinations of first (rows) and second (columns) weaves. It shows inverses in several ways. For example, the opening weave in making the Ten Men figure is to put one's thumb *under* one's near-little-finger string. Any weave that begins with the thumb move just described and results in "resetting the loom" is called an *a weave* in this methodology. The inverse of the foregoing weave is accomplished by putting one's thumbs *over* the string in question and finishing the weave. The "*a inverse*" weave, done immediately after the "a weave," results in the loom's being returned to its beginning state, that is, to where it was before the initial weave. The idea is that any number times its inverse takes the solver back to the number itself, for example, 8 times (1 over 8) returns the solver to 1. With additive inverses, the solver ends at 0, for instance, adding 8 to –8 returns the solver to 0.

Moore (1988) develops a fascinating relationship between string figures and logic, particularly axioms, theorems, and corollaries. A new theorem that is easily derived once a given theorem has been established is called a *corollary* of the given theorem. The same situation prevails in string figures. Many figures are achieved in a very few moves starting with a given figure that has already been constructed. As one example, Ball (1971) describes the method of constructing the pig figure called The Porker. The first instruction for The Porker is "First, make Little Fishes." Ball then develops The Porker in a very few additional moves. By analogy, the elements of the opening position were the "axioms" of the system, Little Fishes was a "theorem" based on the opening position, and The Porker was a "corollary" based on the theorem Little Fishes (Moore 1988).

Students can take away surprising lessons from a string-figure session. In exercise in the Ten Men family, the student performs a series of weaves and then weaves the inverses of those weaves in the opposite order. This process always takes the student back to the beginning loom and shows the power of the inverse weaves. A coauthor's niece used this principle in a mathematics fair to draw the parallel with taking an equation, applying certain operations to it, then applying the inverses of those operations (e.g., multiplying by 3, then dividing by 3), thus ending up with the same equation. A Fortune student reported, "I learned to round off numbers, you know, when to round up or round down." He perhaps got that lesson by analogy from working with strings. Hidden nuggets can be mined in teaching students to work with string figures.

Other Pedagogical Aspects

Several other pedagogical aspects of working with string figures are mentioned by Emerson (1988) or have been noted by the authors in their work with students. Some of these aspects seem self-evident; others deserve elaboration.

- Transformation of raw information into a visible outcome
- Dexterity
- Social skills. Weavers ask for help from and teach one another. An evaluator once said of Murphy's string-figure class that it evidenced the best social interaction of any class that the evaluator has experienced in the multiethnic competitive high school in which the course was taught. The students were concentrating on what was on their hands and were helping one another learn.
- Respect for different cultures. String figures exist in all cultures, all over the world. They show that genius can come from any culture and in an infinite variety of forms.
- Bring order out of chaos. The weaver starts with a dangling strand of string and lifts out a beautiful figure.

One of the beauties of string-figure work is that it instills and teaches all these outcomes in a natural, easy, nonthreatening way. The lessons all arise as natural "corollaries" of doing string figures. Our goal with Learning Hands is to encourage, amplify, and strengthen these qualities by working within families or systems of figures, by encouraging variations and modifications in figures, and by teaching students how to explore the complex and subtle world of string figures systematically. We continue to be excited by Learning Hands' potential in peoples' lives and hope that this promise will promote more extensive use of string figures in the classroom.

LESSON: THE DIAMONDS SYSTEM

The following material is from a book tentatively titled *String Figures by Learning Hands* (Murphy, forthcoming). The book, written in the style of a laboratory manual, is a compilation of lecture notes prepared over the years by coauthor Murphy for his String Figures course, which was designed to teach reluctant mathematics students how to think abstractly within the confines of a rigidly defined system. Algebra, geometry, calculus, and string figures each have a unique set of procedures for keeping order. Within each discipline, students learn the procedures, practice them, abstract them, then learn to apply them to more-complex problems.

The string-figure course examines several basic string-figure families, or *weaving systems*, in great detail. These families include the two-loop Diamonds system (figures related to Jayne's [1962] Osage Two Diamonds), the three-loop Ten Men system (Jayne 1962, pp. 150–56), several three-loop North American Indian Net systems (Jenness's [1924] Fish Net; Jayne's [1962] Many Stars, Owl's Net, and Apache Door), and several four- and five-loop Complex Net systems composed of various Pacific Island figures. The lessons undertake an ordered exploration of ways to vary the parent design within each system. Subsequent lessons focus on how to link various systems to create even more complex designs.

Presented here is the first half of lesson 1, in which students familiarize themselves with a simple two-loop system, Osage Two Diamonds, and learn techniques for adding more diamonds. The second half of the lesson can be found in Murphy (1997); it describes additional techniques for adding complexity and discusses the use of a two-color string for "dissecting" string figures.

Osage Two Diamonds

Osage Two Diamonds, a simple two-loop system, was first described by Jayne (1962, pp. 28–30), who learned it from an Osage Indian. Throughout this article, we use her vocabulary to describe how the string figures are made, sometimes adding terms of our own invention for flavor. To introduce the variations, the steps are presented in three phases: loom phase, weaving phase, and extension phase.

Loom Phase

- Begin with position 1 (fig. 18.1 [a], followed by opening A (fig. 18.1 [b]). Note that in opening A, the left-palmar string is retrieved first. If the right-palmar string is taken up first, the arrangement is called opening B (fig. 18.1 [c]).
- Drop the loop from the thumbs, and pull the hands apart to generate the basic two-loop loom (fig. 18.1 [d]).

Weaving Phase

- With the thumbs, reach away from the body, over three strings, and pick up the fourth string—the far-little-finger string—from underneath. Return the thumbs to their starting position (fig. 18.1 [e]).

Extension Phase

- Use the thumb and index finger of the right hand to pick up the segment of the left-near-index string that lies between the left-index and the left-palmar string. Drop it over the left thumb (fig. 18.1 [f]); then repeat on the right hand, and separate the hands (fig. 18.1 [g]).

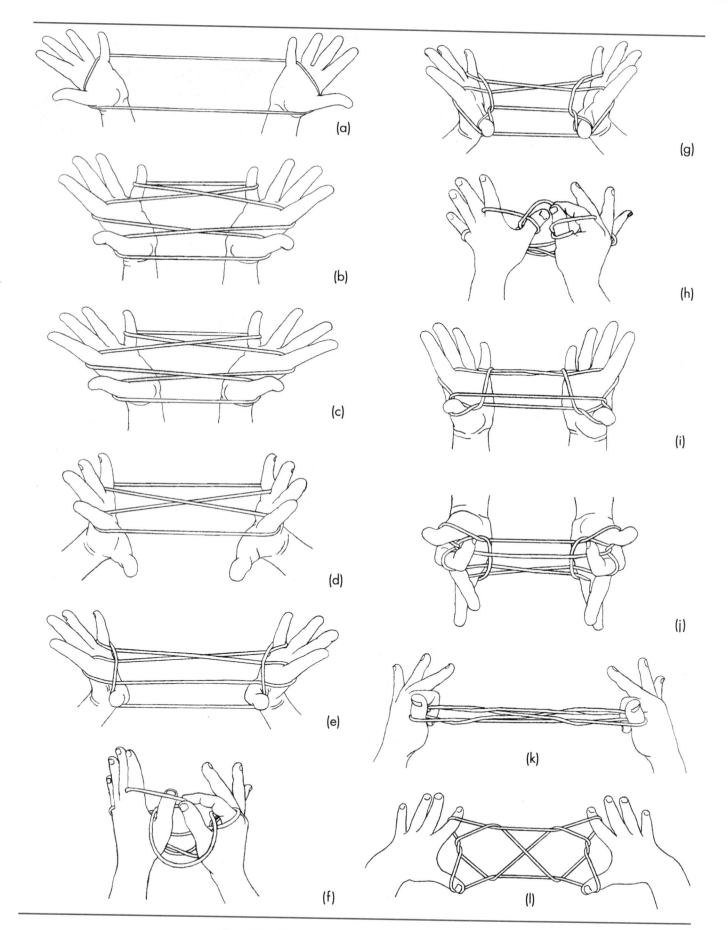

Fig. 18.1. Stages in making Osage Two Diamonds

- With the help of the right hand, _navaho_ the left-thumb loops, that is, lift the lower loop over the upper loop, and over the thumb and release it (fig. 18.1[h]). With the left hand, navaho the right-thumb loops and extend (fig. 18.1[i]). This move is named after the Navaho Indians, the Diné people, who use it frequently in making string figures (Haddon 1912, p. 5).
- Turn the palms toward the body, and note the triangles formed between the thumbs and index fingers. On each hand, place the index finger into this triangle, and press the finger against the palm (fig. 18.1[j]).
- Slide the loop off each little finger, using the right little finger to slide the loop off the left little finger, and vice versa; then separate the hands to absorb the slack (fig. 18.1[k]). Note: The string concealed beneath the bent index finger will become the upper-frame string of the design, so take care that it is not dropped.
- Rotate and open the hands away from the body, as if about to catch a volleyball thrown at shoulder height. As you roll the hands, point the thumbs toward the body, then down, then toward each other; point the index fingers will away from the body and then up as they are straightened. The result is Osage Two Diamonds (fig. 18.1[l]).

For about 95 percent of all students, one or both strings fall off the index fingers during the first few tries. Students benefit from reminders that practice makes perfect and that with practice, their hands will learn to dance smoothly as they form the figure.

Once students have mastered Osage Two Diamonds, they can learn to vary the design by twisting some or all the loops of the two-loop loom (fig. 18.1[d]). The results are One Diamond, Three Diamonds, or Four Diamonds.

One-Diamond Variation

Loom Phase

- Same as Osage Two Diamonds

Weaving Phase

- Bring the hands together, and with the left thumb and index finger, grasp both strings of the right-index-finger loop near the base of the finger. Twist the right index finger toward the body, down, away, and back up to where it started; then let go of the strings held by the left thumb and index finger. In a similar fashion, twist the right little-finger loop a full 360 degrees toward the body (fig. 18.2).
- With the thumbs, reach over three strings, and pick up the fourth string—far-little-finger string—from underneath. Return the thumbs to their starting position.

Extension Phase

- Same as Osage Two Diamonds. The result is One Diamond (fig. 18.3).

Three-Diamonds Variation

Loom Phase

- Same as Osage Two Diamonds

Weaving Phase

- Same as One Diamond, but twist the left-index and left-little-finger loops a full 360 degrees toward the body. (Do not twist the loops on the right index finger and little finger.)

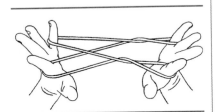

Fig. 18.2. Twisted 2-loop loom

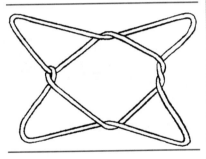

Fig. 18.3. One Diamond

Extension Phase

- Same as Osage Two Diamonds. The result is Three Diamonds (fig. 18.4).

Four-Diamonds Variation

Loom Phase

- Same as Osage Two Diamonds

Weaving Phase

- Same as One Diamond, but in addition to twisting the right-index and right-little-finger loops, also twist the left-index and left-little-finger loops; in other words, twist all four loops a full 360 degrees toward the body.

Extension Phase

Same as Osage Two Diamonds. The result is Four Diamonds (fig. 18.5).

Variations with Opening B

Students can be encouraged to make all the foregoing variations starting with opening B. The results are rather surprising: One Diamond gives Three Diamonds, and with minor differences in string crossings, Three Diamonds gives One Diamond.

Jacob's Ladder

By the end of their first session, most students have acquired the confidence needed to tackle Osage Four Diamonds (Jayne 1962, pp. 24–27), otherwise known as Jacob's Ladder. The teacher might mention that in this four-diamond figure, wraps rather than simple crossings are present where the center two diamonds join the outer two diamonds. Jacob's Ladder is made in the following way.

Loom Phase

- Same as Osage Two Diamonds

Weaving Phase

- Pass the thumbs under all four strings, and bring back the far-little-finger string.
- On each hand, reach over the near-index string with the thumb, pick up the far-index string, and return. Drop the little-finger loops.
- On each hand, reach over the near-index string with the little finger, pick up the far-thumb string, and return. Drop the thumb loops.
- On each hand, reach over both index strings with the thumb, pick up the near-little-finger string, and return.

Extension Phase

- Same as Osage Two Diamonds. The result is shown in figure 18.6.

Next students can be introduced to two repetitive techniques for adding more diamonds to any of the five figures described above. The resulting figures are Storing Two Diamonds and Rastafarian Addition.

Storing Two Diamonds

Loom Phase

- Same as Osage Two Diamonds

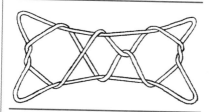

Fig. 18.4. Three Diamonds

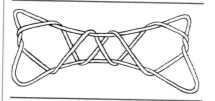

Fig. 18.5. Four Diamonds

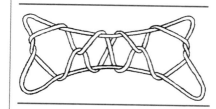

Fig. 18.6. Jacobs Ladder

Weaving Phase

- Pass the thumbs under all four strings, and bring back the far-little-finger string.
- On each hand, reach over the near-index string with the thumb, pick up the far-index string, and return. Drop the little-finger loops.

So far, the method is identical to that for Jacob's Ladder. Next we introduce a modification.

- On each hand, with the little finger, reach over two strings—the near index string and the far-thumb string—then under both the near-thumb strings, and return with the lower-near-thumb string. These moves are a bit awkward, but practice makes perfect. Drop the thumb loops, and pull the hands apart (fig. 18.7).

Two diamonds have just been stored. In fact, moving the hands around a bit permits a glimpse of the two embryonic diamonds near the center. Next complete the weaving and extension phases of Osage Two Diamonds to get "two plus two," or Four Diamonds (fig. 18.8).

To make Six Diamonds (fig. 18.9), store two diamonds, then store two diamonds again, and finish with Osage Two Diamonds, including the extension phase. The technique is additive: the number of diamonds is limited only by the length of the string and the weaver's ability to extend the final figure without the center's collapsing. Six Diamonds can also be made by storing two diamonds, then finishing with Four Diamonds, or by finishing with the entire weaving phase plus the extension phase of Jacob's Ladder.

An odd number of diamonds is also possible. To make Five Diamonds (fig. 18.10), store two diamonds, then finish with weaving phase plus extension phase of Three Diamonds. Surprisingly, the Five Diamonds figure is also obtained by storing two diamonds, then finishing with One Diamond. The only difference between the two designs is the location of the single internal wrap. To make Seven Diamonds, store two diamonds twice, then finish with either Three Diamonds or One Diamond.

Students enjoy exploiting the additive nature of the storing-two-diamonds technique. The ability to build a string figure intelligently with the desired number of diamonds is tremendously satisfying. Students should be encouraged to record the results of their experiments in a table to help them conceptualize the material (see table 18.1).

Fig. 18.7. Storing Two Diamonds

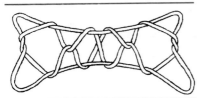

Fig. 18.8. Four Diamonds

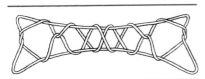

Fig. 18.9 Six Diamonds

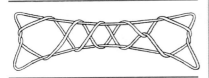

Fig. 18.10. Five Diamonds

Table 18.1
Building on the Storing Two-Diamonds Figure

Start by...	Finish with ...	... and Get
Storing Two Diamonds	Osage Two Diamonds	Four Diamonds
Storing Two Diamonds twice	Osage Two Diamonds	Six Diamonds
Storing Two Diamonds	One Diamond	Five Diamonds
Storing Two Diamonds	Three Diamonds	Five Diamonds
•	•	•
•	•	•
•	•	•

Rastafarian Addition

Another useful technique for increasing the number of diamonds is a continuation move that was taught to Murphy by a Rastafarian from Jamaica on an airline flight. The technique adds another diamond to each end of any finished diamond figure, or to any other figure that has four loops as its extending framework. As in the storing-two-diamonds process, the technique can be applied repeatedly. To illustrate the technique, begin with Jacob's Ladder, then append as follows:

- Pass the thumbs away from the body, under the figure, and at the same time point the fingers toward the body so that the palms face upward. At this point, the diamonds lie parallel to the ground (fig. 18.11[a]). Note the half-twist in each thumb loop. (If you have trouble rotating your hands that fan, you can stop with your palms facing away from the body and your fingers up.)
- On each hand, transfer the half-twisted thumb loop to the little finger, maintaining the half-twist throughout the transfer (fig. 18.11[b]).
- On each hand, pass the thumb over both index strings, pick up the near-little-finger string, and return.

Finish by completing the extension phase of Osage Two Diamonds. The result is Rastafarian Six Diamonds (fig. 18.11[c]), which is similar to Six Diamonds but has unusual wraps between diamonds 1 and 2 and between diamonds 5 and 6. To make Eight Diamonds, simply repeat the process.

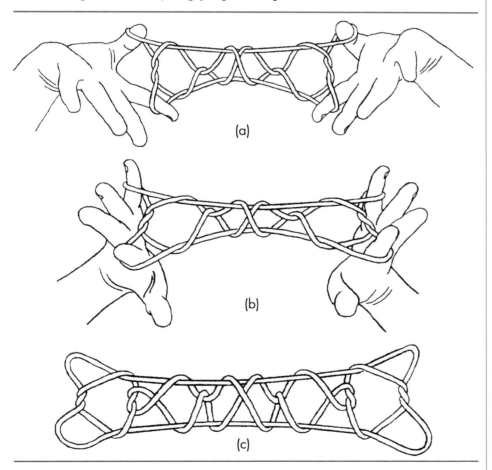

Fig. 18.11. Rastafarian Addition

SUMMARY

Mathematics is the study of the abstraction of patterns, and the peculiar power of mathematics lies in its ability to reduce complexity to controllable procedural patterns. The Diamonds system is a fairly simple suite of figures whose formation teaches a great deal about organizing effort and particularizing differences. Murphy's three-year-old daughter learned the Diamonds system while riding with him to school each morning on the bus. Students push to create their own figure amazingly quickly once they pick up speed and begin to experience success in their searching. A successful experience with string figures can teach students to "learn to learn" mathematics, as well any other intellectual pursuit.

For a more explicitly modern and practical use of string figures, consider the current pedagogical content of modular thinking, in particular, computer programming. A simple example of modularity occurs within the Diamonds system, in which a module of two diamonds can be created and stored, to be added to any of several string figures. This system is a perfect introduction to modern high-level computer languages and their modular approach to complex programming. More-advanced applications of modularity occur when new figures are created using movements borrowed from several different weaving systems. These systems, and the methods of linking them, are explored in subsequent lessons of the Learning Hands series.

REFERENCES

Ball, W. W. Rouse. *An Introduction to String Figures*. Cambridge, U.K.: W. Haffer & Sons, 1920. Reprint, *Fun with String Figures*, New York, Dover Publications, 1971.

Bluestone, M. "Ancient String Patterns Sharpen Skills in Math." *New York Times*, 19 June 1986, p. C10.

Emerson, Larry. "NA'AT'LO—Dine String Games as an Educational Process." Unpublished manuscript, 1988.

Hapgood, David. Personal conversation, New York, spring, 1995.

Jayne, Caroline F. *String Figures and How to Make Them*. 1906. Reprint, New York: Dover, Publications, 1962.

Jenness, D. "Eskimo String Figures." *Report of the Canadian Arctic Expedition*, 1913–1918, vol. 13, part B. 1924.

Métraux, A. "Ethnology of Easter Island." *Bishop Museum Bulletin* 160 (1940).

———. translated from the French by Michael Bullock. *Easter Island—a Stone-Age Civilization of the Pacific*. New York: Oxford University Press, 1957.

Moore, Charles G. "The Implication of String Figures for American Indian Mathematics Education." *Journal of American Indian Education* 28 (1) (1988):16–26.

Murphy, James R. "Using String Figures to Teach Math Skills—Part 1: The Diamonds System." *Bulletin of the International String Figure Association* 4 (1997): 56–74.

———. *String Figures by Learning Hands*. Forthcoming.

Storer, Tom. Personal communication, December 1994.

Stevenson, Gelvin. "What Learning Hands Teach: An Exploration of the Psychological, Emotional, and Conceptual Impact of Making String Figures." *Bulletin of the International String Figure Association* 2 (1995): 6–19.

Walton, Kendall L. *The New York Times* Book Review, 18 June 1995.

Ratio, Proportion, Similarity, and Navajo Students

19

Claudia Giamati

This article examines several areas of mathematics in which cultural differences between the Navajo culture and Western culture manifest themselves. These differences are due primarily to fundamental differences in language structure and in part to philosophical beliefs. The article examines how conceptual understandings in mathematics are influenced by these factors. Finally, it provides examples that illustrate how the teacher can learn to communicate with, and teach mathematics to, Navajo students while maintaining the integrity of the students' cultural mode of coming to know the concepts and that of the mathematical structures being examined.

BACKGROUND

Ascher (1992) has pointed out that language has a deep influence on a person's view of mathematical relationships. The Navajo language is one of the few languages that has more verbs than nouns (Pinxten 1983; Witherspoon 1977). Pinxten (1983, 1985, 1991) has identified the differences that exist between Navajo and Western cultures and has examined the dynamic action-oriented language that influences the Navajo's way of acquiring knowledge. The process of creating or doing something is often more important than the finished product (Rhodes 1994; Ascher 1992). Children learn to describe an object by its function and its geometrical characteristics, as well as by its location in "space" (Begaye 1995), because of the structure of the Navajo language.

Numerous studies and assessment tools have shown that Navajo students have great difficulty with ratio and proportion and, thus, with fractions. The action of dividing things into parts of wholes is not encountered in the Navajo language or culture (Rhodes 1994; Pinxten 1983). Unfortunately, some educators would have us believe that simply changing "apples and oranges" to "horses and sheep" in a primary school student's word problem will take account of her or his cultural differences and make the mathematics problem relevant to the Native American student. Activities of this type, for example, asking students to count the number of sheep in a herd and subtract the number of males or brown sheep from the total number of sheep, may be perceived by educators as a culturally relevant, real-life application. However, if counting sheep in a herd is not an activity that members of the culture customarily perform (Smallcanyon 1995), then this mathematical activity is not actually a real-life application. A Navajo shepherd may count the number of sheep in the herd, but the essential features of the herd do not rely on the number of sheep present. The herd is seen as a whole unit with a distinctive behavior. Each individual sheep is also conceived as a complete entity with a distinctive behavior. Therefore, if twenty sheep are in the herd, then one of those sheep is seen not as one-twentieth of the herd but as the individual with brown fleece or with a nick in its ear.

In his resistance to the act of dividing, the shepherd might be compared to a sculptor who, having obtained a piece of granite from a mountain, would never say that the piece is one ten-thousandth of the mountain (Weiland 1995). The sculptor sees a new individual piece of rock that will soon take on an individual purpose as it is sculpted. For similar philosophical reasons, fractions, ratios, and proportions represent crucial stumbling blocks in the Navajo child's mathematical education.

However, whereas cultural differences account for difficulty with certain mathematical topics in the Navajo culture, they also account for a different set of strengths. Ample evidence shows that the Navajo have a rich intuitive understanding of geometry and space (Pinxten 1983, 1985, 1991; Giamati and Weiland 1995). One of the focuses of the geometry of the Navajo is the art of rug making. Their rug patterns have been widely used as a basis for judgment about the Navajo's intuitive understanding of geometry and symmetry (e.g., Zaslavsky [1991]). Although this topic is appropriate as a model for a discussion of symmetry, it only scratches the surface of Navajo students' intuitive geometric understanding based on their cultural experiences. Two art forms that can illuminate more deeply what we might expect students to understand intuitively about geometry are sand painting and basket weaving. An examination of such art reveals much more complex patterns than are usually encountered in mainstream Western culture.

This strength was further revealed when an open-ended patterning exercise based on Gilliland's (1991) work was given to Navajo students (Giamati and Weiland 1995). The Navajo students resisted the sort of linear planar patterning that students from many cultures produce. Rather, they chose to construct patterns that revealed an abundant use of asymmetry and manipulations of objects in space. When figures were repeated, they were not repetitive congruent shapes but, rather, were similar objects that were always proportionately smaller or larger than the preceding one. These notions illuminated the fact that the students' perceptions of many relations were very different from what we expected. However, we also found ways to explain topics in mathematics through approaches based on these perceptions. What follows are several examples that illustrate an alternative approach to discussing ratio, proportion, and similarity with Navajo students.

These alternative approaches were used with fifty-six Native American students, two of whom were Hopi and the others, Navajo. Of the fifty-six students, only nine were raised with English as their first language. The others were bilingual, with Navajo and Hopi as their primary languages. All the students lived on the Navajo or Hopi reservations. Their school, Grey Hills High School in Tuba City, Arizona, was located on the Navajo reservation. The students were all ninth and tenth graders who were taking an introductory algebra course. They had all been exposed to ratio, proportion, and equivalent fractions before, without much success.

GEOMETRIC SEQUENCES, RATIOS, AND EQUIVALENT FRACTIONS

To devise the examples that follow, we relied on three elements from the Navajo culture:

1. The dynamic, action-oriented focus of the Navajo language

2. The rich, intuitive understanding of geometry possessed by the students

3. The importance of the natural and temporal world in Navajo culture

The most important change for us, as members of Western culture, was to look for a dynamic action or process that could be used to illustrate ratio and proportion. The students led us to use growth, rather than division, as a natural vehicle to teach ratio and proportion. Changes in size can be examined in terms of growing and shrinking, both important verbs in Navajo culture, instead of according to the standard practice of looking for a part of the whole. For example, corn, an important staple of Navajo culture and legend, is small as it sprouts and grows bigger until it is harvested. This growth was used to introduce ratio and proportion in terms of the object's growth. The students were asked to imagine a stalk of corn when it first sprouts, then to guess how tall it would be when fully mature. They were then asked how tall it would be after one week, and so on, until a diagram like the one in figure 19.1 was produced.

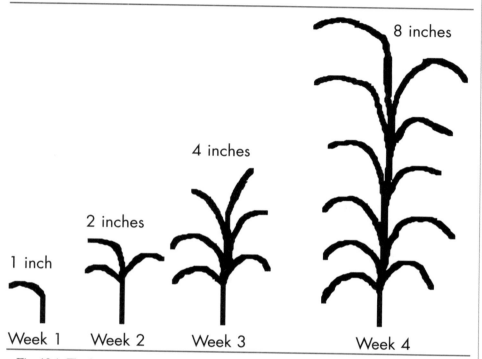

Fig. 19.1. The hypothetical heights of corn at the beginning of the first four weeks of growth, illustrating a geometric sequence

At this point in the lesson, the types of questions asked by the teacher about the heights of the corn were more influential than her constructions of the visual representation. The focus of the question was on the action, the verb. Such questions as "What rate did the corn 'grow by'?" highlighted the action involved in the corn's growth. The teacher proceeded to ask the students to decide what rate the corn "grew by" from the first sprout to the second week. Once the student identified that the corn's height was multiplied by 4, she then asked them, "Since we don't have a height for week 1, how could we use our diagram and some mathematics to figure out how tall the corn would have been then?" She then allowed the students to create their own diagrams of similar situations and to demonstrate how their objects grew and shrank. The students were given the option of working in groups or as individuals.

The students then created visual representations of geometric sequences, but the teacher did not want to introduce new vocabulary by naming them as such. Instead, she demonstrated the notion of growth ratio by comparing the sizes of consecutive figures. "Let's look at our corn example, and compare the heights at week 1 and week 2," she said. "What is the growth ratio?" The students quickly noticed that it is two times larger.

T: We write this ratio as

$$\frac{2}{1} = 2.$$

Now what about week 2 and week 3—how could we use our method of recording the growth ratio?

Students [giving the inevitable answer]:

$$\frac{2}{4}.$$

T: Is this the same type of method we used in the first example?

S1:   No, the larger number is on top.

T: Good, so we should write

$$\frac{4}{2}$$

to record our ratio.

S1:   Yes, that's it.

T: Can I write

$$\frac{4}{2} = 2$$

like we did before?

S1:   Yes.

T: Can I write

$$\frac{4}{8} = 2?$$

S2:   No, it's not the same.

T: What is different?

S1:   The larger number must stay on top.

T: Why?

S3:   It looks like division, and it's grown two times.

T: What if we were shrinking instead, going backwards?

S2:   Then it wouldn't be a growth … ratio.

T: What do you think we could call it?

S3:   Maybe a shrink, but would it be two times?

T: No, what would it be?

S1:

$$\frac{1}{2}.$$

The teacher then used a different growth-and-shrink ratio to highlight the difference between growth and shrink ratios. The notion of 1/2 is used in the Navajo culture, but other fractions are not integrally used. Several more examples, with a variety of ratios, allowed students to see the relation and calculations involved in shrinking. Once this foundation had been established, at least two different mathematical topics built on this same example and helped students learn the use for fractions in a dynamic application instead of as objects to be manipulated.

SIMILARITY

Our research showed that Navajo students have a clear intuitive, albeit naive, understanding of similarity and projective geometry. *Projective geometry* involves knowing exactly where the shadow of an object would lie and what shape it would possess if the location of the sun was given. Ideas from projective geometry gave us a way to use the corn example to build a conceptual foundation for similarity.

The teacher used the idea of shadows cast by the sun and asked the students which would have the longer shadow, the tallest stalk of corn or the smallest. A student astutely inquired where the sun was located, and the teacher produced the diagram in figure 19.2. It should be noted that since the sun moves, an artificial light source should be used. The students quickly remarked that the smaller corn plant would have the smaller shadow. The teacher then asked, "If we knew the small plant was 1 inch tall, the larger one was 4 inches tall, and the shadow of the smaller plant was 2 inches, could we find out how big the longer shadow is?"

S1: It should be 8.

T: Good, how do you know that?

S1: The little plant is 1, and its shadow is stretched to 2, so the 4 is stretched to 8.

T: Right; now, what if the plant was 3 inches tall?

S2: The shadow is 6

T: OK, now let's use this notion to talk about a mathematical relation called *similarity*. What we're interested in is growing and shrinking again and finding the growth and shrink ratios.

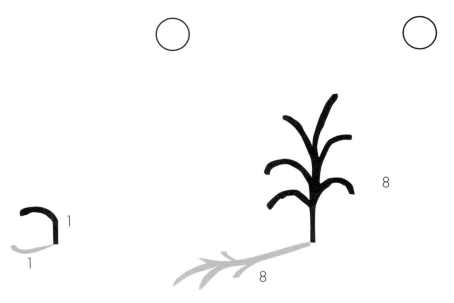

Fig. 19.2. The diagram used to compare the lengths of shadows

The teacher then constructed triangles by using the corn and lining up the shadows so that they met at the vertex of a triangle, as shown in figure 19.3. The requirements for similarity, angle congruence, and sides in proportion were easily grasped by the students because they had a strong conceptual foundation on which to build. In general, the only problem that Navajo students had with similarity was the skill-oriented task of checking for proportional sides or finding the lengths of missing sides. This weakness came from the fact that no relevant conceptual link had been made between their experience and similarity. Once such a link had been established, the nuances and details of similarity were then easily tackled by the students. They were successful in solving problems that had once been stumbling blocks for them.

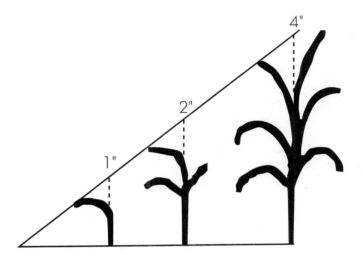

Figure 19.3: The construction of similar triangles on corn plants and their shadows

DISCUSSION

All the students who helped us learn to view and explain these topics in an alternative way had previously been exposed to the topics. Although they benefited from these alternative approaches, it remains to be seen whether experiencing the topics in this way will be as beneficial as their earlier exposure. But what we have illustrated here is not an exhaustive method of teaching ratio and proportion or similarity. Readers should understand that many details have been omitted in our discussion of the topic.

We can see how certain cultural differences can influence the way we communicate mathematical ideas effectively or ineffectively. Were we to use the standard conceptual part-whole model that has been proved effective with Western children, Navajo students would be left confused and further torn between two worlds. In fact, the use of separation is so repugnant to the Navajo that the teacher could at times lose their attention entirely. One such situation would be looking at rug patterns and asking students to visualize taking apart the pattern. Instead, the teacher must ask the students to look within the pattern for whatever symmetry or shapes are to be investigated.

Although Navajo students have exhibited weaknesses in mathematics due to cultural interference, they have shown strengths, as well. Their visualization skills far surpass those of most Western children. Their understanding of space and time is also excellent. The fundamental reasons for these differences is their language. Thus, a teacher who is devising ways to discuss the attributes of a topic should focus on these strengths and use them to shore up the students' weaknesses. If we want to teach children in such a way that maintains both the integrity of their culture and that of the mathematical structures, we first must learn the fundamentals of that culture from our students.

When the teacher shows a genuine interest in teaching a given subject, students of any culture respond to this interest. Although Navajo students will not often question the teacher, they will answer questions and help the teacher understand their perceptions of concepts. The teacher can then use this knowledge to help students understand the mathematical concept while maintaining both cultural and mathematical integrity.

BIBLIOGRAPHY

Ascher, Marcia. *Ethnomathematics: A Multicultural View of Mathematical Ideas.* New York: W. H. Freeman & Co., 1992.

Begaye, M. Personal conversation and interview, 1995.

Bradley, Claudette. "Issues in Mathematics Education for Native Americans and Directions for Research." *Journal for Research in Mathematics Education* 15 (March 1984): 96–106.

Congdon-Martin, Doug. *The Navajo Art of Sand Painting.* Westchester, Penn.: Schiffler Publishing, 1990.

Davison, David. "Mathematics." In *Teaching American Indian Students*, edited by John Reyhner, pp. 241–50. Norman, Okla.: University of Oklahoma Press, 1992.

Giamati, Claudia, and M. Weiland. "Mathematics as a Verb and Not a Noun." Paper presented at AERA-Sig ME Research Pre-Session to the NCTM Annual Meeting, Boston, April 1995.

Gilliland, K. *Patterns of Culture.* Unpublished manuscript, 1991.

Hadfield, O., J. Martin, and S. Wooden. "Mathematics Anxiety and Learning Style of the Navajo Middle School Student." *School Science and Mathematics* 92, no. 4 (1992): 171–75.

National Council of Teachers of Mathematics (NCTM). *Curriculum and Evaluation Standards for School Mathematics.* Reston, Va.: NCTM, 1989.

Orban-Szontagh, Madeleine. *Southwestern Indian Designs.* Mineola, N.Y.: Dover Publications, 1992.

Pinxten, Rik. "Knowledge as a Cultural Phenomenon." In *Language Acquisition and Learning: Essays in Educational Pragmatics*, edited by F. Vandamme and E. Spoelders. New York: Plenum Press, 1985.

———. "Geometry Education and Culture." In *Learning and Instruction*, pp. 217–27. Great Britain: Pergamon Press, 1991.

Pinxten, Rik, Ingrid Van Dooren, and Frank Harvey. *The Anthropology of Space: Explorations into the Natural Philosophy and Semantics of the Navajo.* Philadelphia, Pa.: University of Pennsylvania Press, 1983.

Rhodes, Richard W. *Nurturing Learning in Native American Students.* Hotevilla, Ariz.: Northland Graphics, 1994.

Smallcanyon, T. Personal conversation and interview, 1995.

Watson, H. "Investigating the Social Foundations of Mathematics: Natural Number in Culturally Diverse Forms of Life." *Social Studies of Science* 20, no. 1 (1990): 283–312.

Weiland, Brad. Personal conversation and interview, 1995.

Witherspoon, Gary. *Language and Art in the Navajo Universe.* Ann Arbor, Mich.: University of Michigan Press, 1977.

Zaslavsky, Claudia. "World Cultures in the Mathematics Classroom." *For the Learning of Mathematics* 11, no. 2 (1991): 32–36.

Symmetry and Beadwork Patterns of Wisconsin Woodland Indians

20

Bernadette Berken

Kim Nishimoto

M athematics is the study of patterns. Patterns abound in nature, and the patterns are frequently adapted and used by humans in a variety of artistic and creative endeavors. This chapter reinforces the connections between the patterns used in strip or border designs by Wisconsin Woodland Indians and the mathematics behind these patterns. It focuses on patterns of symmetry in strip or border designs.

FOUR TYPES OF RIGID MOTION IN THE PLANE

Mathematicians can describe the balance or symmetry of a figure by using the idea of rigid motion, or *isometry*. A rigid motion is a transformation in space or in a plane in which the original figure and its image are congruent. Four kinds of rigid motions are possible in the plane: (1) reflection through a line, (2) rotation, (3) translation, and (4) glide reflection.

The simplest isometry is reflection across a line. The resulting symmetry is called *bilateral symmetry* or *mirror symmetry*. A figure with bilateral symmetry looks the same on both sides of a line except that the two sides of the figure are mirror images of each other. Figure 20.1a shows an example of a figure with a vertical line of symmetry. Figure 20.1b shows an example of a figure with a horizontal line of symmetry. Figure 20.1c shows an example of a figure with both vertical and horizontal lines of symmetry.

Courtesy of Neville Public Museum of Brown County
Fig. 20.1a. A figure with a vertical line of symmetry

Courtesy of the State Historical Society of Wisconsin
Fig. 20.1b. A figure with a horizontal line of symmetry

Courtesy of Neville Public Museum of Brown County
Fig. 20.1c. A figure with both a vertical and a horizontal line of symmetry

A second, equally simple type of rigid motion is rotation. A figure that demonstrates *rotational symmetry* has a center about which it can be rotated by a certain angle without changing its overall appearance. Figure 20.2 shows an example of a figure with rotational symmetry. For a strip pattern, the only possible angle rotation is through an angle of 180 degrees. Otherwise, the strip pattern will not be preserved.

Translation is the third type of plane isometry. In this rigid motion, the figure slides a certain distance in a certain direction. Only an infinite figure can be translated without undergoing a change in its appearance. Figure 20.3 shows a pattern with *translational symmetry*. The reader will notice that the figure repeats itself indefinitely along the strip.

From the personal collection of
Bernadette Berken
Fig. 20.2. A figure with rotational symmetry

Courtesy of the State Historical Society of Wisconsin
Fig. 20.3. A figure with translational symmetry

The fourth rigid motion, glide reflection, is not as simple or familiar as the first three. It is really a combination of two motions: a glide or translation some particular distance along a line followed by a reflection across that same line. The pattern that a person's footprints leave in the snow is an example of a simple pattern that exhibits *glide-reflection symmetry*. Figure 20.4 shows an example of glide-reflection symmetry.

Courtesy of Neville Public Museum of
Brown County
Fig. 20.4. A figure generated by a glide reflection

Activity 1, adapted from Berken (1996) and found at the end of this chapter, involves students in creating a stencil pattern and then using the pattern to design paper "border strips" that illustrate each of the four types of symmetry motions in the plane.

CLASSIFICATION OF STRIP-PATTERN TYPES

Because every strip pattern can be constructed from one or more of these four kinds of rigid motion applied to a single design unit, a simple system for classifying patterns easily evolves. This method of classification has been used for a long time and, in fact, was first developed by crystallographers who wanted to classify the three-dimensional patterns that were found in crystals.

The classification of strip patterns in the plane is a simple process requiring no complex tools. Only seven one-dimensional strip-pattern types result from the various possible rigid motions. The seven types can be classified using a four-character classification. The first character for a strip pattern is always a "p." The second character indicates whether the pattern exhibits vertical symmetry: it is an "m" if the pattern has a vertical line of symmetry; otherwise the it is a "1." The third character gives information about the horizontal or glide symmetry of the pattern. It is an "m" if the pattern exhibits a horizontal line of symmetry; it is an "a" if the strip pattern has a glide reflection but not a horizontal line of symmetry. Otherwise it is a "1." The fourth character is a "2" if the pattern has a point of 180-degree rotational symmetry; otherwise it is a "1."

These four-character classification codes give complete information about the mathematical symmetry elements of any conceivable strip pattern. Consider the two beadwork patterns in figure 20.5. Although very different to look at, both are mathematically equivalent in terms of the symmetry elements that they possess. Both can be classified as pmm2.

Courtesy of Neville Public Museum of Brown County

(a)

Courtesy of the State Historical Society of Wisconsin

(b)

Fig. 20.5. Two patterns that can be classified as pmm2.

A simple flowchart can be easily used to classify any strip pattern with this four-character code. The flowchart in figure 20.6 is an adaptation of one that is presented and more completely described in Washburn and Crowe (1988). A copy of the flowchart should be distributed to students to assist them in completing the worksheet "Exploring Symmetry of Rigid Motions That Create Strip Patterns," which is adapted from Berken (1996). Answers to the worksheet can be found at the end of this article.

Activities 2–4, adapted from Berken (1996) and found at the end of this article, engage students in hands-on experiences that encourage creative thinking while reinforcing students' learning of the seven one-dimensional strip-pattern types. In activity 2, students use the initials of their names to create patterns that reveal the seven strip-pattern types. In activity 3, students color a section of graph paper to produce a design and then apply one of the rigid motions to their design to make a strip pattern that illustrates one of the seven basic types of such patterns. Activity 4 involves students in making a loom-beaded bracelet that incorporates the graph-paper strip pattern that they designed in activity 3.

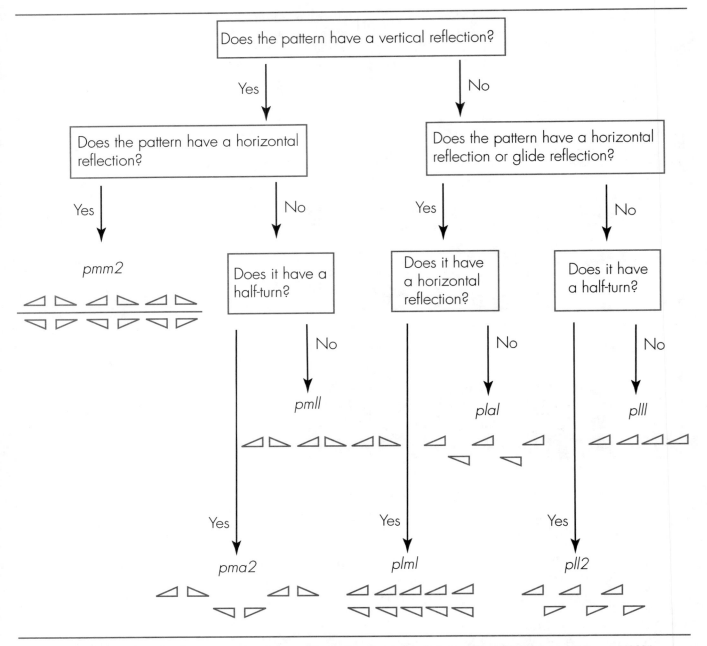

Fig. 20.6. Flowchart for the seven one-color, one-dimensional patterns (adapted from Washburn and Crowe [1988])

Worksheet

EXPLORING SYMMETRY OF RIGID MOTIONS
THAT CREATE STRIP PATTERNS

1. For each of the following strip patterns, identify the rigid motions that preserve the pattern:

(a)   M M M M M M M M

   Pattern code: _____

(b)   E E E E E E E E E

   Pattern code: _____

(c)   X X X X X X X X

   Pattern code: _____

(d)   O O O O O O O O

   Pattern code: _____

(e)   N N N N N N N N

   Pattern code: _____

(f)   Z Z Z Z Z Z Z Z

   Pattern code: _____

(g)   A A A A A A A A A

   Pattern code: _____

2. Use the standard notation of crystallographers described in the flowchart (p. 212) to identify each pattern below according to the type of symmetry that it represents.

(a)

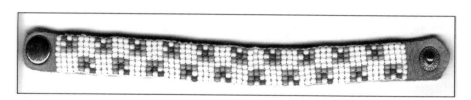

Pattern code: _____

(b)

Pattern code: _____

From *Perspectives on Indigenous People of North America*

Worksheet—*Continued*

(c)

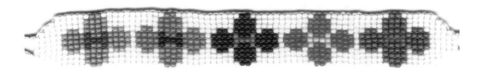

Pattern code: _____

(d)

Pattern code: _____

(e)

Pattern code: _____

(f)

Pattern code: _____

(g)

Pattern code: _____

(h)

Pattern code: _____

(i)

Pattern code: _____

Worksheet—*Continued*

3. Again using the notation of crystallographers described in the flowchart (page 212), identify each pattern below according to the type of symmetry that it represents.

(a)

Menominee belt, courtesy of the State Historical Society of Wisconsin

Pattern code: _____

(b)

Menominee sash (belt), courtesy of the State Historical Society of Wisconsin

Pattern code: _____

(c)

Chippewa sash, courtesy of the State Historical Society of Wisconsin

Pattern code: _____

(d)

Chippewa sash, courtesy of the State Historical Society of Wisconsin

Pattern code: _____

(e)

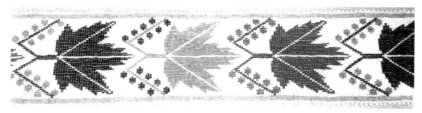

Winnebago sash, courtesy of the State Historical Society of Wisconsin

Pattern code: _____

From *Perspectives on Indigenous People of North America*

Worksheet—*Continued*

(f)

Menominee bandolier band, courtesy of Neville Public Museum of Brown County

Pattern code: _____

(g)

Chippewa beaded strap on beaded bag, courtesy of Neville Public Museum of Brown County

Pattern code: _____

(h)

Menominee sash, courtesy of Neville Public Museum of Brown County

Pattern code: _____

Activity 1: Stencil Designs for Reflection, Rotation, Translation, and Glide-Reflection Symmetry

Objective

Students will draw designs demonstrating each of the four symmetry motions.

Materials Needed

Manila folders or cardstock-weight paper

Pencils

Craft knives

Paper

Activity 2: Initial Strip Patterns

Objective

Students will use their initials to create a motif to which they will apply rigid motions to create strip patterns to illustrate the seven basic strip designs.

Consider the designs created by Tammy V. and Theo. Both show that even a simple motif involving only the initial or initials of a name can form interesting strip patterns when a simple isometry is used.

 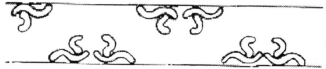

Materials Needed

Paper

Pens, pencils, markers, or paint

Calculator paper tape

Activity 3: Strip Patterns

Objective

Students will create a simple design on graph paper. They will apply a rigid motion to the design to create a strip pattern illustrating one of the seven basic strip designs.

Materials Needed

Graph-paper strips

Markers

Activity 4: Loom-Beaded Bracelets Using Strip Patterns

Objective

Students will use the strip patterns that they created in activity 3 to make a loom-beaded bracelet.

Materials Needed:

Scissors

Beading needles

Nylon thread, size B

Cotton-covered polyester thread, medium weight

Beads, size 10

Beeswax

Bead loom *

*Bead looms can be purchased at some craft stores, or a simple wooden bead loom can be constructed as shown in figure 20.7.

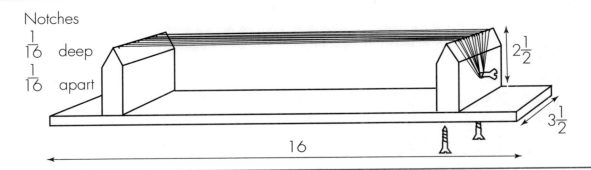

Fig. 20.7. Dimensions for constructing a simple wooden bead loom

Activity 1

STENCIL DESIGNS FOR REFLECTION, ROTATION, TRANSLATION, AND GLIDE-REFLECTION SYMMETRY

Procedure

1. With a pencil, draw a simple shape on a three-inch-by-three-inch square of manila folder or other cardstock, making sure that you leave a small one-half- to one-inch border around your design. Choose a shape of design that does not have any inherent symmetry element.

2. Cut around the stencil pattern with a craft knife.

3. Cut the stencil out of the manila folder by following the lines of the one-inch border.

4. Use the homemade stencil to create design strips that demonstrate each of the four symmetry motions. Draw your designs in the spaces provided below.

   (a) Reflection

   (b) Rotation

   (c) Translation

   (d) Glide reflection

Activity 2

INITIAL STRIP PATTERNS

Procedure

1. Create a simple design using your initial or initials.

2. Use your initial motif to create strip patterns on calculator paper tape to illustrate each of the following rigid motions:

   (a) Translation

   (b) Glide reflection

   (c) 180-degree rotation

   (d) Vertical reflection

   (e) Horizontal reflection

3. Label each of your strips with the type of rigid motion illustrated.

4. Label each strip to give its four-character strip-pattern code.

From *Perspectives on Indigenous People of North America*

Activity 3

STRIP PATTERNS

Procedure

1. Create a design by coloring in different squares of the small graph-paper section below with markers. You must color in whole squares, not portions of squares, on the graph paper. Your design should be five to eleven squares wide so that it can be used later for a simple loom-beading activity. Cut out the colored design.

2. Apply one of the rigid motions to your cutout design to create a strip pattern on the large section of graph paper below.

3. Label your completed strip pattern with the type of rigid motion that it illustrates.

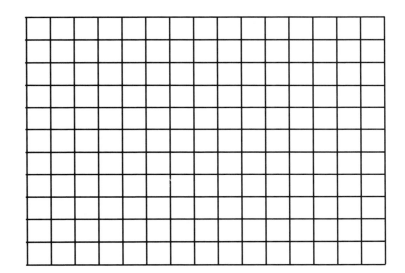

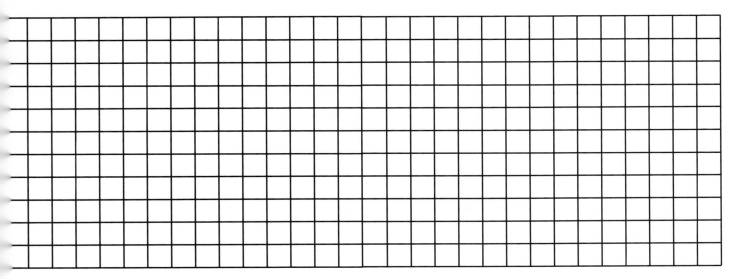

From *Perspectives on Indigenous People of North America*

Activity 4

LOOM-BEADED BRACELETS USING STRIP PATTERNS

Procedure

Stringing the Loom:

1. Tie the end of the cotton thread to the screw on one side of the loom.

2. Bring the thread up into the notch in the loom. Extend the thread the length of the loom to the notch in the other side. Pull the thread taut, and wrap it around the screw on the other side.

3. Continue stringing the loom in this manner, extending the loom thread first in one direction and then in the other, passing through corresponding notches and wrapping the loom thread around the screw until you have the total number of loom threads needed for the design. Since the beads are positioned between the loom threads, the number of loom threads must be one more than the number of beads in the width of the design. For example, if the design is eleven beads wide, twelve loom threads will be needed.

4. Double the loom thread on both of the outermost threads, and knot the thread around the screw. The doubled threads give the finished project more strength.

5. Run the beeswax over the loom threads. This step strengthens and protects the loom threads.

Beading on the Loom:

1. Thread the beading needle with the nylon thread, and run the thread through the beeswax.

2. Tie the end of the nylon beading thread to one of the outermost loom threads, and press the excess thread along the loom thread in the direction of the beading.

3. To start beading the design, string the appropriate number of beads on the needle to the beading thread. Refer to the strip pattern to determine the color and number of beads to string.

4. Slide the beads down the nylon thread to the knot. Pass the needle and thread carrying the strung beads under the loom threads. With your index finger, press the beads up between the proper loom threads. While holding the beads in place, pull the beading thread taut, take it up over the outside loom thread, and pass the threaded needle back through the beads above the loom threads, securing the row of beads. See the illustration below.

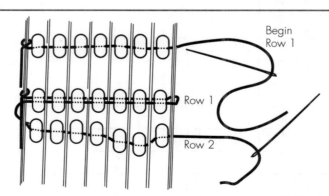

Activity 4—Continued

5. Bead the remaining rows in the same manner, referring to the strip pattern for the number and color of the beads to string.

6. When the beading thread becomes short and is sufficient to do only a few more rows, thread the beading thread back through the beads in the previous four finished rows. Cut off the excess thread.

7. Cut and wax another piece of nylon beading thread. Thread the beading needle.

8. Start the new beading thread by passing the threaded needle through the last four rows of beads completed. Make sure that you exit the last row by threading over the outermost loom thread. Continue loom beading as described in step 4.

9. When the beaded strip is completed, thread the beading thread back through the last four rows of beads. Cut off the excess thread.

10. To remove the beaded strip project from the loom, cut the loom threads at the screw.

11. On each end of the strip, divide the loom threads into groups of three and braid them. Finish by tying a knot at the end of the each braid. The bracelet can be worn by tying the two braided ends together. Alternatively, the beaded strip can be sewn to a leather strip to which snaps or ties can be added for closure.

From *Perspectives on Indigenous People of North America*

ANSWERS TO WORKSHEET

1. (a) translation, vertical reflection
   (b) translation, glide reflection, horizontal reflection
   (c) translation, glide reflection, vertical reflection, horizontal reflection, 180-degree rotation
   (d) translation, glide reflection, vertical reflection, horizontal reflection, 180-degree rotation
   (e) translation, 180-degree rotation
   (f) translation, 180-degree rotation
   (g) translation, vertical reflection

2. (a) p112   (b) pmm2
   (c) pmm2   (d) pm11
   (e) p1a1   (f) p111
   (g) pmm2   (h) pma2
   (i) p1m1

3. (a) pma2   (b) pm11
   (c) pmm2   (d) pmm2
   (e) p1m1   (f) p1a1
   (g) p112   (h) p111

REFERENCES

Berken, Bernadette. *Symmetry Patterns of the Wisconsin Woodland Indians: A Curriculum Resource.* De Pere, Wis.: Saint Norbert College, 1996.

Washburn, D. K., and D. W. Crowe. *Symmetries of Culture: Theory and Practice of Plane Pattern Analysis.* Seattle and London: University of Washington Press, 1988.

Counting on Tradition

Iñupiaq Numbers in the School Setting

21

Wm. Clark Bartley

This chapter sets forth the role of an inquiry-based model in developing a numeral-notation system for the indigenous oral counting tradition among the Alaskan Iñupiat Eskimos. It then goes on to describe the subsequent invention by Alaska Native schoolchildren of new Kaktovik Iñupiaq numerals, which led to a revolutionary approach to mathematics. The innovation all began as a simple enrichment activity in a middle school mathematics class in Kaktovik, an Iñupiat village on a remote island in the Arctic Ocean, just off the coast of northern Alaska. The premise behind this chapter is that shared inquiry, together with indigenous Native American number sense, can promote the establishment in the classroom of a "sociomathematical community" of Native student-mathematicians. The approach, which was basically constructivist, has led not only to sound mathematical concepts and a richer educational experience for the students but also to significant discoveries about mathematical systems. In the Iñupiat experience, the introduction into the classroom of the traditional numbering system went well beyond simple ethnomathematical exploration; it gave birth to new numerals, offered students a new way to exercise number sense, and paved the way to different and exciting ways of approaching computation.

The history of the development of Iñupiaq mathematics in the schools of the North Slope Borough School District has been a kind of spontaneous explosion of energy from within one small school on a remote island in the Arctic Ocean. The phenomenon was both unplanned and unexpected. It is a story of discovery by young Iñupiat, which has brought with it the energy to catapult Iñupiaq mathematics into international attention. Already, in the short time since its inception, it has been reported in the press from Anchorage to New York and from Washington, D.C., to Paris, and it has been taught to students in classrooms across the North Slope of Alaska—from young children in the Early Childhood Education Immersion Program in Barrow to adults in college classes, as well as to young Eskimo schoolchildren in Yup'ik-speaking areas in southwestern Alaska. Students from Point Hope to Barter Island have actually been discovering how to do mathematics in a different way that is based on the genius of their traditional Iñupiaq counting system.

The North Slope Borough is a unique geographic entity. It covers an area of nearly 90,000 square miles, about the same as the combined areas of the states of Ohio and Pennsylvania, with a population of around 7,000 persons, predominantly Iñupiat Eskimo. As a borough, it is a regional political division of the state of Alaska, but it is the only borough in Alaska whose boundaries are coterminous with those of a *regional corporation*, that is, a larger tribal unit. The economic, political, and tribal center of the North Slope is Barrow, with about 4,000 inhabitants.

The rest of the population is spread out among seven villages, Kaktovik being the furthest to the east. Kaktovik is located on Barter Island, which lies just off-shore from the Arctic National Wildlife Refuge. It is one of the larger barrier islands and is the only inhabited island in the Beaufort Sea, having a population of more than 200 people. The Canadian border is less than 90 miles to the east, whereas about 600 miles separate Kaktovik from Point Hope, the westernmost village of the North Slope Borough.

Theoretical Perspective

When this project began, it truly had no theoretical direction; in fact, when it began, it was not really a "project" at all. It started as a short enrichment activity and ended up becoming a fortuitous grand detour—one that shows the promise of becoming a main highway to intellectual autonomy in mathematics for future Iñupiat schoolchildren.

My training, like that of many teachers assigned to teach mathematics, has been largely outside mathematics, and I had never even heard of constructivism. My first degree was in anthropology, and I also have a background in Latin, Greek, and Sanskrit and a master's degree in linguistics. Having studied the classics, I have always been attracted to the Socratic method of shared inquiry. In class discussions of literature, I pose questions to which I do not myself know the answer, so that true shared inquiry can actually take place. In mathematics, however, I was arrogant enough to believe that I already knew the answers. Even so, I always tried to frame discussion questions that would lead the students into thinking.

Not surprisingly, when my students and I began to explore the base-twenty counting system of the Iñupiaq language, with its subbase of five, I fancied that I had the answers; I assumed that the Iñupiaq counting system was just another base. I was wrong. I did not realize the significance of a subbase, but my students, fortunately, were as perceptive as they were persistent. Without knowing it, we were embarking on a great "constructivist" expedition into mathematics.

DEVELOPING NUMERALS FOR IÑUPIAQ

In early September 1994 at Harold Kaveolook School, middle school students were exploring base-two numbers in their mathematics class. Some students mentioned that Iñupiaq, their Eskimo dialect, has a base-twenty system. They then decided to try to write the Eskimo numbers with regular Arabic numerals but found that not enough symbols were available adequately to represent the Eskimo numbers to their satisfaction.

Creating the Symbols

The students addressed their dilemma by creating ten extra symbols, which proved to be difficult to remember and some of which were so fanciful that an inordinate amount of time was needed to write them. Since nine students composed the entire middle school, we were able to involve all of them in a class discussion. After discussing the problems, the students tried a number of approaches but were not satisfied with the outcomes. We held another major brainstorming session, but this time, not to create numerals but instead to list the qualities that our ideal numeral system should have. Here are a few of the criteria that the students themselves established:

A. The symbols should be easy to remember.

    1. A relationship should be evident between the symbols and their meanings.

    2. The symbols should reflect the way that one counts in Iñupiaq.

B. The symbols should be easy to write.

   1. If possible, each symbol should be written without lifting the pencil from the paper.

   2. The shapes should be easy to write quickly.

C. The symbols should be distinct from the Arabic numerals so that no confusion would result between the two systems.

D. The symbols should be pleasing to look at.

The class took these criteria and tried to come up with a set of symbols that reflected the Iñupiaq counting system. Intuitively, they realized that their counting system was built in groups of five. To give the reader an idea of the system with which the students were working, figure 21.1 presents the Iñupiaq number-words up to twenty.

atausiq	malġuk	piŋasut	sisamat
1	2	3	4
tallimat	itchaksrat	tallimat malġuk	tallimat piŋasut	quliŋŋuġutaiḷaq
5	6	7	8	9
qulit	qulit atausiq	qulit malġuk	qulit piŋasut	akimiaġutaiḷaq
10	11	12	13	14
akimiaq	akimiaq atausiq	akimiaq malġuk	akimiaq piŋasut	iñuiññaġutaiḷaq
15	16	17	18	19
iñuiññaq				
20				

Fig. 21.1. Iñupiaq number-words up to twenty

Ultimately the students came up with an idea that was conceptually simple and that admirably reflected the Iñupiaq oral counting system. However, they realized that they would also need a symbol for zero to complete the system. They understood that since twenty was the base of their system, twenty should be written with the symbols for one and for zero. No traditional word existed for zero in Iñupiaq, but the students nevertheless needed to create a symbol for it. Aware that counting in Iñupiaq had traditionally been done on fingers and toes, one of the sixth-grade girls suggested that the symbol for the zero should look like "crossed arms," indicating that nothing is being counted. After fine-tuning, the students' new notation system comprised the written numerals shown in figure 21.2.

Computing with the New Symbols

Once the students were satisfied with their new numerals, they wanted to use them for computation. They began to do simple addition and subtraction problems with them. To their amazement, they discovered that the new symbols had many distinct advantages. Adding and subtracting with them were easier than with Arabic numerals. Often the symbols almost gave the students the answer. In adding, the students could easily visualize combining two numerals into a single figure, which is the sum of both; for example:

$$\vee + \vee = W \qquad > + \diagup = \gtrless \qquad \overrightarrow{\vee} + \overrightarrow{\vee} = \overline{\overline{W}}$$

The eye easily combines the two parts into the sum.

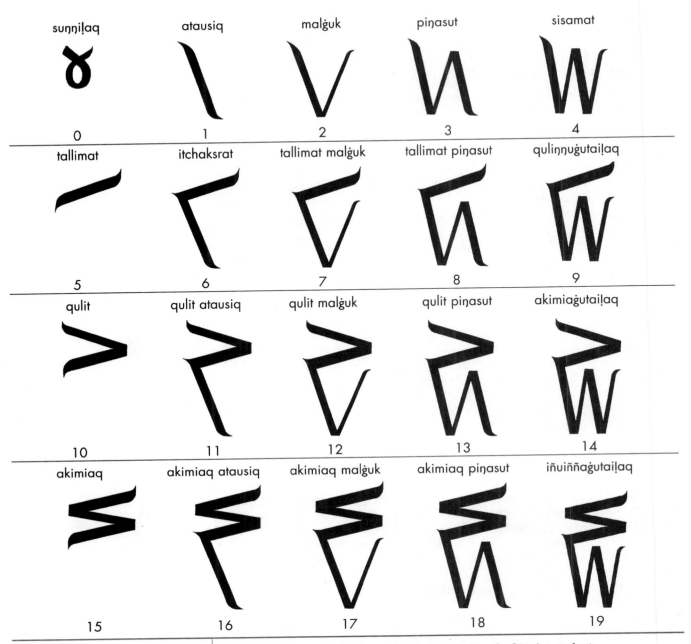

Fig. 21.2. The numerals invented by the Iñupiat students

Subtraction is even easier. One simply finds the shape of the subtrahend in the minuend and deletes it mentally; the answer contains the strokes that are left. The less computationally proficient students stopped counting on their fingers to add and subtract. Instead, they began converting problems into their new base-twenty symbols to find solutions. The new numeral system actually allowed a more visual approach to solving problems, not an insignificant fact.

Research suggests that the Iñupiat possess a higher-than-normal level of visual-perceptual skills, no doubt an adaptive mechanism of the Iñupiat culture in response to requirements for survival in the harsh Arctic environment, where permanent landmarks are often lacking and where shifting snows complicate orientation. In her comparative study of visual memory among Iñupiat and Caucasian schoolchildren, Judith Kleinfeld concluded that teaching strategies in Iñupiaq village schools should build on these visual-cognitive strengths by symbolizing concepts in progressively more abstract images ("Visual Memory in Village Eskimo and Urban Caucasian Students," *Arctic* 24 (2) (1971): 132–38). The new numeral system capitalized on these inherent strengths. Perhaps instinctively, the Iñupiat schoolchildren of Kaktovik invented a numeral system that is quite visually symbolic, both of the Iñupiaq number words and of their conceptual numerical value.

Teaching the New Numerals to Younger Pupils

The students were excited; they asked whether they could teach their numerals to the younger pupils in the school. The teachers agreed, and pupils in the lower grades began to learn the numerals, as well. This experience was the beginning of a new phase in the middle school students' understanding of the implications of the Iñupiaq counting system and the notation system that they had invented. When the middle schoolers had to assume the role of tutors, they found that the younger students did not always write the new numerals in the way they, the inventors, conceived them. Perhaps because of interference from the Arabic numerals and even the alphabet, the younger pupils were inclined to write all the numerals so that they touched the bottom line when written on lined paper. That approached lowered the numerals ⌐, ⌐, and ⌐ from the higher position that the middle school students thought they should occupy.

Back in the classroom, we discussed the situation. I suggested that perhaps the vertical position of a symbol did not really matter; the numeral was still recognizable. I even suggested that maybe the younger pupils would feel more as if they, too, were making a contribution if my students adopted the way that the younger pupils wrote the numerals. I almost won the day, but some of the students thought that they would be conceding an important point. As it turned out later, they were right.

Creating Number Frames

The middle schoolers next turned their attention to brainstorming methods to help the younger pupils observe the writing conventions that they had established. The students developed their own terminology to teach the numerals. They began to call the upper strokes, representing groups of five, "bars"; they called the lower strokes, representing units, "lines." Finally they developed a device that they called "number frames" to help the younger pupils write the Kaktovik numerals "correctly." See figure 21.3. The frames that the middle schoolers created were essentially upright golden rectangles intersected horizontally by a line segment, resulting in a square at the bottom and, at the top, a smaller golden rectangle on its side. The golden rectangles that they drew by hand became somewhat distorted when we recreated them on the computer to make worksheets for the younger pupils, but from that time forward, the students began talking in terms of "upper sector" and "lower sector," as represented in the new numeral frames.

Each student in the class was assigned to teach the new symbols to one or two younger pupils. To keep the Kaktovik numerals distinct from the Arabic numerals, the Iñupiaq number-words were always used to name Kaktovik numerals, whereas English was used for the Arabic numerals. After each session, the students came back to the classroom to discuss their experiences. When they met and considered a specific problem that the younger pupils were having, they discussed the factor that they thought caused the problem and then brainstormed pedagogical solutions. The opportunity to teach younger students seemed to make them acutely sensitive to attributes of the numeral system that they had created. They would soon realize that they had actually invented a two-dimensional place-value system.

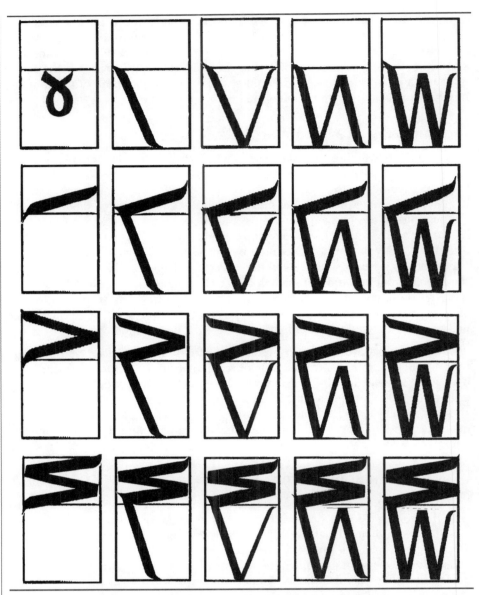

Fig. 21.3. Student-invented "number frames" to encourage the correct position of written Kaktovik numerals

FROM VISUAL TO TACTILE NUMBER SYMBOLS

Generally, the students enjoyed the challenge of converting decimal numbers into their new base-twenty Kaktovik numerals. As they tried to convert increasingly larger numbers, they found that conversion was easier using counters with a place value. This idea was then developed into a form of a base-twenty abacus. The students discussed the ideal structure of their abacus and decided that it would have three beads above the read-bar and four beads below the read-bar. See figure 21.4.

The students secured beads from the art teacher, experimented, and finally built abacuses in the school shop. At first they used the abacus primarily to convert large numbers from the decimal system into the Iñupiaq system. They quickly realized that their base-twenty abacus represents values in a way similar to their new numerals. Later they began experimenting in using it for computation. They found the abacus to be a simple and useful tool, not only for

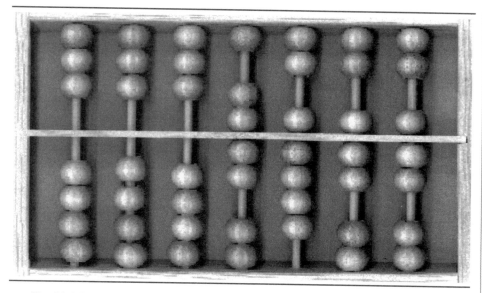

Fig. 21.4. Iñupiaq abacus showing the number ⵣⵯⵣⵣ, or 99747, the local Zip code

converting numbers from base ten to base twenty but also for adding, subtracting, multiplying, and dividing. The base-twenty abacus has become an important component in mathematics education involving the Kaktovik numerals.

The abacus also exemplifies the two-dimensional aspect of the students' system. The students began to recognize that their "sectors" were actually a secondary, vertical place-value system and that the orientation of the bars and lines in their numerals was merely a writing convention to help distinguish the place value of the upper and lower places. This orientation of the strokes is important because a "zero" placeholder is not written in the lower sector if a value is represented in the upper sector. Gradually it became apparent that regrouping takes place on two levels. Five units are regrouped upward as one group of five, and four groups of five are regrouped to the lower left as one group of twenty. See figure 21.5.

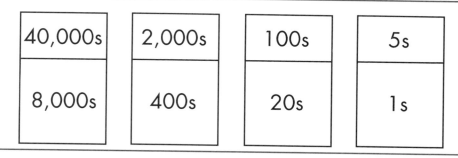

Fig. 21.5. Students' invented number-frame "sectors" representing a vertical place-value system in which regrouping takes place on two levels

To demonstrate this regrouping concept to the younger pupils, the middle schoolers rethought the Cuisenaire models that they had constructed previously. Initially, they had built only "rods" of twenty cubes, "tiles" of 400 cubes, and a "large cube" of 8,000 smaller cubes. At this point, they recognized the need to build "strips" that were 5 cubes long to represent groups of five, "paddles" that were 5 cubes by 20 cubes to represent groups of 100, and "slabs" that were 5 cubes by 20 by 20 to represent groups of 2,000. To teach addition, and particularly the concept of regrouping fives, they also used Popsicle sticks and rubber bands. Out of the appropriate number of sticks, younger pupils could then form the shapes of the actual Kaktovik numerals for addends. Single sticks were used as "lines" in the lower sector, or ones place, and bundles of five sticks were used as "bars" in the upper sector, or fives place (see fig. 21.6). Rubber bands were often used to hold bundles together. With these aids, even the youngest pupils could easily regroup in their heads as they worked problems from a printed page of base-twenty problems.

| 3 | + | 4 | = | 7 |

Fig. 21.6. Popsicle sticks were used to form Kaktovik numerals and to teach regrouping by fives.

IÑUPIAQ MATHEMATICS

Iñupiaq mathematics, to the extent that it now exists as a scholastic discipline, was born as a twin on the heels of the Kaktovik numerals. As the middle school students began to perform more and more mathematical operations with their numerals, they discovered that their symbols were powerful enough to be manipulated as symbols. The symbols themselves seemed to function as a kind of graphic mathematics manipulative.

Visual Procedure for Division

When the students began to experiment with division, they started out by using the same familiar algorithm that they had always used for dividing base-ten numbers. That approach was made even more difficult by the fact that none of the students knew the multiplication facts up to 20 times 20. However, soon a few students noticed that the process could be simplified because of the visual nature of the numerals that they had invented. For example, the numeral ⚊ (15) looks essentially like the numeral ʌ (3) turned on its side and raised into the upper sector. So when visually dividing ⚊ (15) by ʌ (3), they find ʌ just once, but because it is found on its side in the upper sector, the quotient is also written on its side in the upper sector. The numeral one (\) written in the upper sector actually becomes the numeric symbol for a five (⌐).

Next the students made a startling discovery. Even two- and three-digit divisors could be divided visually into very large dividends, as long as no regrouping was necessary. Soon the students figured out how other long-division problems could be solved visually, almost as though they were short-division problems requiring no knowledge of multiplication facts. No need arose for multiplying and subtracting. Examples of this visual procedure follow:

$$\frac{30,561}{61} = 501 \qquad\qquad \frac{45,349,226}{2,826} = 16,401$$

A visual examination of the first problem reveals the constituent strokes of the divisor numerals ⋀\ can be found once in the lower sector of ⋀⋾, the first two numerals of the dividend. The divisor is then found once, oriented on its side, as ⋍ in the upper sector of the two middle numerals ⋾⋀ of the dividend. Therefore, in the quotient, the second numeral is written in the upper sector and is oriented horizontally, representing a "five" in the twenties place. Finally, the constituent strokes of ⋀\ can be found once again in the lower sector of ⋀\, the last two numerals of the dividend. The students discovered the useful tactic of tracing with a colored pencil the constituent strokes of the divisor as they found those in the dividend, to keep track of what strokes they had used. In the second problem, the constituent strokes of the divisor, ⋁\⋾, can be seen twice in ⋿⋿⋀, the first three numerals of the dividend. The rest of the problem can be worked out in a similar fashion with the aid of a colored pencil.

Extending the Subbase into a Two-Dimensional Place-Value System

To simplify regrouping, the students extended the subbase vertically beyond the spectrum band of spoken Iñupiaq number words. Although this extension was purely theoretical at first, the students later found evidence for it in the Cuisenaire models that they had built to illustrate their number system. Figure 21.7 is a chart illustrating the extension of the subbase into a more fully two-dimensional place-value system and showing the value of each position in the grid. The shaded area represents the spectrum-band of "potential" spoken number names. The term *potential* is used because names for all the fractional numbers, as well as very large numbers, did not exist in the tradition of the Iñupiaq language; elders and Native educators needed to agree on a system for naming these quantities. The double bars extend the point that separates whole numbers, in the units place, from the vigesimal, or base-twenty, fractions to their right and from the quintesimal, or base-five, fractions below.

1,000,000	50,000	2,500	125	6.25	.3125
200,000	10,000	500	25	1.25	.0625
40,000	2,000	100	5	.25	.0125
8,000	400	20	1	.05	.0025
1,600	80	4	.2	.01	.0005
320	16	.8	.04	.002	.0001

Fig. 21.7. Chart illustrating the extension of the subbase five into a two-dimensional place-value system

Division can be carried out visually by extending the subbase place-value system upward just one additional place and by introducing two inverse procedures, which we call "fragmenting" and "defragmenting." These two procedures are "shortcuts" that students discovered to simplify regrouping between the extended subbase and the spoken spectrum. Still later, they discovered that extending the subbase place-value system vertically can simplify multiplication, and even extracting square roots. They also discovered other new procedures, or "tricks," to simplify their visual algorithms. Quite frequently, they discovered shortcuts in mathematics that are simply impossible with the corresponding Arabic numerals.

Another particularly significant step was students' discovery of the importance of what we now call "complements." One extremely visually oriented learner noticed that ⋿ (19) is the most "complete" numeral; it has a full set of strokes in both sectors. Every other number is lacking some strokes. The "missing strokes," when combined, comprise the complement of the numeral (e.g., \ is the complement of ⊼, ⹀ is the complement of w, and ▽ is the complement of ⊽.) When doing regrouping for subtraction, he observed that the difference was always related to the complement of the numeral in the subtrahend. Over time, we have discovered several other applications for the concept of "complements."

FORMAL ADOPTION

In spring 1995, the North Slope Borough Board of Education invited the students from Kaveolook School to attend the school board's work session to present and explain their invention. Those who attended that presentation were impressed with the exciting educational possibilities opened up by this system. The system is a direct reflection of the way that one counts in Iñupiaq. The underlying genius of the Iñupiaq language has been crystallized in the new numerals, making them useful for practical purposes. New life has been breathed into the ancient, traditional Iñupiaq counting system by contemporary students who have come to regard themselves as mathematical thinkers in their own right. In the following year, the Commission on Iñupiaq History, Language, and Culture adopted the numerals to represent the numbers in the Iñupiaq language.

In Kaktovik many people take pride in the fact that the Iñupiaq way of counting is on the verge of entering the twenty-first century triumphantly as a vital contribution to education. At least in Kaktovik, people now recognize that the whole world is watching what is happening with their number system.

As the 1995–1996 school year began, the ECE (Early Childhood Education) immersion class in Barrow and the Iñupiaq language classes in Wainwright and Point Lay began introducing the numerals into the classrooms. Teachers in other grades at the elementary school, the middle school, and even the high school in Barrow, Alaska, began introducing the system to their students. Ilisagvig, the local community college, added an "Iñupiaq Mathematics" course to its catalog and offered classes through compressed video to introduce the numerals and their use to students across the North Slope.

By this time, a great deal had been discovered about the practical potential of the Kaktovik numerals, and the students and their teacher had collected a great deal of material about other Arctic and Native American counting systems. In a letter that they received, written by Robert Petersen from the University of Greenland, they learned that the Eskimo counting system is technically called a *pentavigesimal* system, meaning that it has a subbase of five and a base of twenty. Until recently, many people had given little thought to the counting system developed and used by the ancestors of contemporary North Slope Iñupiat, but now that very system is generating national and international interest. Dozens of articles have been published about the Kaktovik Iñupiaq numerals in local Alaskan newspapers, magazines, and newsletters, as well as in the national and international press. Radio and television reports have also been broadcast about them.

In 1997, after all the students who invented the Kaktovik numerals had moved on into high school, the principal agreed to have teachers devote half the time in the middle school mathematics classes to instruction in the pentavigesimal system, using the Kaktovik numerals. That first year, the middle school students' scores on the California Achievement Test in mathematics dramatically increased. The previous year, the mean score for these students was below the twentieth percentile. After a year of studying both base-ten and base-twenty mathematics simultaneously, their mean score roses to significantly above the national average.

CONCLUSION

Curiosity and ethnic pride led the students in Kaktovik to forge themselves into a group with a common mathematical purpose. Their tenacity has rewarded them with great dividends. They discovered much more than a new notational system. They discovered that mathematics is all around them, just waiting to be "discovered." They discovered a whole new approach to computation. They tasted the excitement of independent discovery. They discovered that their Native pentavigesimal counting system behaves quite differently from the traditional decimal system because of the fact that the former has a subbase.

Equally important, these students have also discovered on their own that the decimal system has no inherent superiority, nor does their own traditional base-twenty system have any inherent inferiority. And through this discovery, they have realized another advantage. Just as the person who has mastered and can speak two languages well surely has an advantage over a person who knows only one language, so the person who masters and becomes adept in manipulating two different number systems very likely has an advantage over one who understands only one number system. Although one can count in "scores" in English (e.g., four score and seven), this system is archaic and unfamiliar to twentieth-century Americans, so some real ingenuity might be required to replicate this experience for students who speak only English. However, in a day when mathematics skills are becoming increasingly important in the job market, perhaps all people who speak a language with a base-twenty counting system should seriously consider including instruction in base-twenty mathematics as an integral part of their school mathematics curricula.

The Aztec Number System, Algebra, and Ethnomathematics

22

Luis Ortiz-Franco

Mathematics education originated when human beings began to quantify the phenomena in their lives. Although the process of counting was the same for different groups of people around the world, the symbols by which they represented specific quantities varied according to particular cultural conventions. Thus, African, Aztec, Babylonian, Chinese, Mayan, and other cultural groups wrote numbers in many different ways.

Today, even within a single society, different subgroups (some defined by profession, such as accountants, physicists, engineers, mathematicians, and chemists), view and manipulate mathematical quantities differently. The study of the particular way in which specific cultural or ethnic groups go about the task of mathematizing their environment is called "ethnomathematics" (D'Ambrosio 1985a, p. 2; 1985b, p. 44). Ethnomathematics, which includes the cultural and historical dimensions of mathematical knowledge, provides a conceptual framework for integrating multiculturalism in mathematics education throughout the curriculum instead of treating it as a special academic topic. When we integrate multiculturalism into our classroom teaching of mathematics, we provide students with opportunities to view mathematics as a dynamic and vibrant universal human activity practiced by diverse cultural groups. By exposing students to ethnomathematics, we increase their appreciation for the mathematical achievement of other cultures.

Today, cultural minorities (African Americans, American Indians, and Latinos) are severely underrepresented in mathematics-based careers. For example, even though these three groups collectively constituted 23 percent of the U.S. population in 1993, only 12 percent of bachelor's degrees in science and engineering were awarded to members of the three groups in that year (National Science Foundation 1996, pp. 2–4). Ethnomathematics can be used as a tool to motivate these disenfranchised students to pursue their study of mathematics. As teachers, we can provide such students with relevant mathematical experiences by integrating into the curriculum mathematical topics from their own cultures. In so doing, we maximize the possibilities for improving their attitude toward mathematics at the same time that we are improving their mathematical skills. Hence, ethnomathematics is a medium through which we can have a positive impact on the affective and cognitive domains of students who are underrepresented in mathematics.

As the twenty-first century opens, the demographics of our mathematics classrooms are changing, becoming increasingly diverse in cultures, ethnic groups, and languages. NCTM's *Curriculum and Evaluation Standards for*

Portions of this article were prepared while the author was a visiting scholar at the Linguistic Minority Research Institute (LMRI) of the University of California at Santa Barbara in the fall of 1994. Other portions of the work were supported by a summer research grant in 1994 and a travel grant in summer 1996, both from Chapman University. The opinions expressed are those of the author and do not necessarily reflect the views of LMRI or Chapman University.

School Mathematics (NCTM 1989) (*Curriculum Standards*) and *Professional Standards for Teaching Mathematics* (NCTM 1991) (*Teaching Standards*) acknowledged this reality in the closing years of the previous century as they addressed the notion of connecting mathematics to other disciplines and cultures.

Teaching Standards recommends that teachers emphasize connections between mathematics and other disciplines (p. 89). Moreover, it advises teachers to consider students' cultural backgrounds in designing and selecting mathematical tasks (pp. 26, 115). In particular, it suggests that teachers guide students in exploring different numeration systems from the history of mathematics so that students will appreciate mathematics as a changing and evolving domain to which many cultural groups have contributed (p. 26). In addition to recognizing the pedagogical virtues of a multicultural approach in the teaching of mathematics, *Teaching Standards* acknowledges the importance of using multicultural perspectives in achieving the important objective of motivating students from underrepresented groups to study more mathematics (p. 146). Ethnomathematics is instrumental in implementing these standards.

Ethnomathematics is also vital to implementing *Curriculum Standards*. The fourth standard for the three clusters of grades—K–4, 5–8, and 9–12—recommends that connections be made among different mathematical topics as well as among mathematics and other disciplines (pp. 35, 84, 146). Many topics in school mathematics can be connected with ancient civilizations' number systems, history, and geography in introducing new concepts and helping students develop proficiency in particular mathematical areas. Therefore, ethnomathematics can facilitate the achievement of two important objectives in the teaching of mathematics. It can establish a multicultural context for mathematics knowledge and skills, and it can assist students in making connections among other cultures and disciplines.

For example, the number systems of the Aztecs and the Egyptians can illustrate concepts related to the addition and multiplication of polynomials in algebra. When discussing those two cultures, we can use maps to demonstrate their territorial influence, and we can identify the historical period when they thrived. Both cultures had nonpositional number systems that did not use zero, and both wrote their numerals according to an additive principle by which symbols could be written in any order. This convention makes numerals in these systems unambiguous because their symbols uniquely determine the value represented.

The addition of quantities in such number systems emphasizes collecting like terms in a process similar to that followed in the addition of polynomials in algebra. The multiplication of two numbers in those systems of numeration can be accomplished by introducing parentheses or brackets around the factors, applying the distributive property, and collecting like terms in a process similar to that followed in the multiplication of polynomials in algebra. Nelson (1993, p. 50) makes general comments concerning the addition of Egyptian numerals, and Joseph (1991, p. 64) illustrates a method for their multiplication.

This article presents a classroom application of ethnomathematics that uses an American Indian number system—the system of the Aztecs—in the teaching of algebra and makes connections among culture, history, geography, and mathematics. The article provides two lessons for the classroom as well as a series of recommendations for teachers. The first lesson outlines the pre-Columbian number system of the Aztecs, and the second lesson uses that system to introduce the concept of addition and multiplication of polynomials.

The objective of this lesson is to introduce the Aztec number system and use it to explore connections among mathematics, history, and geography.

Number Signs

Historical evidence suggests that the Aztecs employed a vigesimal (base twenty) number system (Cajori 1928, p. 41; Vaillant 1962, p. 171). Figure 22.1 shows Aztec symbols for some important numbers in such a system. Sources agree that the Aztecs used a dot to represent the quantity 1. Vaillant (p. 171) and Payne and Closs (1993, pp. 228–29) suggest the possibility that a finger was also used occasionally to represent the number 1. To write numbers from 1 to 4, one simply drew as many dots or fingers as necessary. The present discussion is limited to using dots for the units.

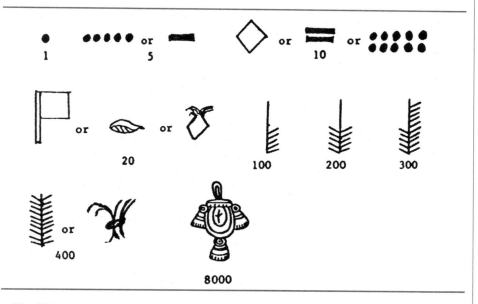

Fig. 22.1. Aztec symbols for 1, 5, 10, 20, 100, 200, 300, 400, and 8000, including different ways of writing 5, 10, 20, and 400

The number 5, which all sources indicate was represented by five dots, could also be expressed as a single bar, according to an indigenous informant, David Vasquez, an Aztec descendant whose first language is Nahuatl, the language of the Aztecs, and who grew up in the Aztec community of Tlalmotolo in Puebla, Mexico. To write numbers from 6 to 9, one simply wrote as many dots as required or used a combination of one bar and the appropriate number of dots. Some sources say that the Aztecs represented the number 10 by a lozenge or rhombus; other sources make no mention of those symbols. Vasquez says that 10 could also be represented by two bars and that three bars could represent 15, although other sources do not mention those conventions. If Vasquez's assertions are correct, one could use any one of the following conventions to represent the numbers from 1 to 19: (1) dots alone, (2) dots in combination with bars, or (3) dots in combination with bars and lozenges. Symbols can be written in any order when a number requires two or more of them.

The number 20 was represented by a flag, but Vasquez adds that two other symbols—a shell-like sign and a type of vase with feathers or grass stems protruding from it—were also used to represent that quantity. The largest number symbols shown in figure 22.1 are those for 400 and 8000. Four hundred was represented by a sign like a fir or a feather with twenty hairs or barbs, each of which represented 20. Vasquez reports that 400 was also represented by a sign resembling a bundle of grass or plants. Our sources agree that 8000 was written as a purse or baglike sign.

Payne and Closs (1993, pp. 216, 229) assert that a flag was repeated as many times as necessary to represent multiples of 20 that are less than 400. Other sources indicate that 100 could be represented either by five flags or by five barbs on the fir or feather, and that 200 and 300 were represented by ten and fifteen barbs, respectively, on the fir or feather.

The conclusion that the Aztecs' number system was vigesimal is based on the observations that $400 = 20 \times 20 = 20^2$ and that $8000 = 20 \times 20 \times 20 = 20^3$. However, we do not know what symbols (if any) the Aztecs may have used for the numbers 160,000 $(= 20^4)$ or 3,200,000 $(= 20^5)$ or higher powers of 20. Ascher (1991, pp. 7–8) also comments on Aztec numbers (which she refers to as Nahuatl numbers), noting that those numerals have a cyclical pattern based on 20.

Writing Numerals

The Aztecs' number system is not positional; numerical representations are unambiguous and do not include zero. (The Aztecs did, however, have many names for the concept of zero as the beginning of the numbers, including *teolot*, which means "god of knowledge about the beginning of numbers." In the Aztec system, a number's symbols are additive and can be written in any order. Consequently, there are many ways to write numerals that use three or more different symbols, depending on the order in which one arranges the symbols. For example, the number 8375 can be written in more than 240 different ways, depending on which symbols are chosen to represent 20 (a flag, a vase, or a shell), 10 (a lozenge or two bars), and 5 (a bar or five dots), and in which order the symbols are written. Figure 22.2 shows four different ways of writing 8375. The last two representations show permutations in the order of each other's symbols.

The variety of possibilities for writing Aztec numbers that require three or more symbols presents teachers with an opportunity to introduce a multicultural perspective into classroom discussions of permutations and combinations in discrete mathematics. When teachers discuss the writing of Aztec numerals, they should point out similarities and differences between the Aztec and the Roman number systems. This approach will allow them to present the writing of the Aztec numerals from a constructivist perspective. Teachers should, of course, make it clear that Aztec numerals often involve symbols that enlarge them, by addition, but never involve symbols that reduce them, by subtraction, as some Roman numerals do (for example, IV and IX).

Origins of the Aztec Number System

Teachers should use two maps when turning their students' attention to the connections among Aztec mathematics, history, and geography. One map should show Europe and the New World, and another should show Mesoamerica so that teachers can identify the geographical region where the Aztec civilization thrived. (A map of Mesoamerica can be found in Stuart [1981], and a larger map of the same region can be obtained from the National Geographic Society in Washington, D.C.)

When Columbus accidentally arrived in the New World in 1492, the dominant civilizations in these newfound lands were the Aztecs of Mexico and the Incas of Peru. The Aztec civilization developed from a wandering, desperate band calling itself *Mexica*, which first appeared in the Valley of Mexico about

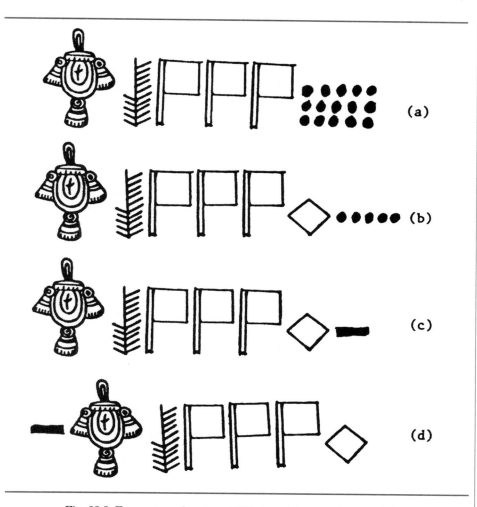

Fig. 22.2. Four ways of writing 8375 with Aztec number symbols

A.D.. 1200. Historical evidence indicates that the Aztec civilization established its principal base in Tenochtitlan, now Mexico City, in July 1325. At its height in 1519—just before the Spanish conquest—the empire ruled over millions of subjects and embraced lands from the Gulf of Mexico to the Pacific Ocean and from northern central Mexico to the present Guatemalan border.

As the Aztecs expanded their territorial influence, they usually left defeated rulers in place as long as they paid tribute. The conquerors allowed the vanquished cultures to retain their own customs, religion, and social institutions. Moreover, the Aztecs frequently adopted cultural and religious practices from the conquered civilizations, often adapting those borrowings to enrich their own cultural and religious institutions.

This practice may have played a role in the use of the dot for the number 1, the bar for 5, and the shell for 20. These symbols were the three elements of the number system often called Mayan, which was actually created by the Olmecs around 1200 B.C. The Olmec civilization—the oldest one known to pre-Columbian Mesoamerican history—disappeared around 700 B.C. In the Olmec system, the shell-like symbol represented 0, whereas the Aztecs used it to represent 20. Whether the Aztecs created the other number signs or whether they were developed by other Mesoamerican civilizations is unclear. However, no evidence indicates that other Mesoamerican civilizations used the same number system as the Aztecs.

Figures 22.1 and 22.2 show that the Aztec number system is cumbersome to use for writing large numbers or adding or subtracting large quantities because of all the symbols that one must draw to represent the numbers. However, recognizing connections between the processes of writing multiple-symbol Aztec numerals, adding them, and subtracting them and the processes of writing and manipulating multiple-term expressions in algebraic notation can be very beneficial to beginning algebra students.

Ideas for the Classroom and for Homework

Teachers can ask students to answer the following questions either in class or at home. The purpose of the questions is to strengthen the students' understanding of the Aztec number system, to require them to interpret historical events, and to foster their critical thinking. (Answers are provided at the end of the article.)

1. Write 1200 as an Aztec numeral to stand for the year A.D. 1200. In the history of the Aztecs, what is the significance of this year?
2. Write 1325 at least two different ways in Aztec symbols. Consider the year A.D. 1325 in Aztec history. What is its significance?
3. Write 1 492 using Aztec symbols, and let the numeral stand for the year A.D. 1492. If there were other people (such as the Aztecs and the Incas) living in the New World when Columbus arrived, is it accurate to say that Columbus "discovered" the New World? Explain your thinking.
4. Write 1519 at least three different ways using Aztec symbols. Consider the year A.D. 1519. What is its significance in Aztec history and world history?

Teachers are encouraged to create additional questions and exercises congruent to the objectives of this unit.

LESSON 2: ALGEBRA AND THE AZTEC NUMBER SYSTEM

The objective of this lesson is to use the Aztec number system to reinforce students' learning of particular algebra skills. Writing and working with Aztec numbers at the same time that they are learning to write and work with algebraic expressions can help students see links between "old" and "new" at the notational and the algorithmic levels. The fact that Aztec numerals are very visual makes them especially valuable for illustrating some basic algebraic concepts. With respect to notation, the "old" Aztec numerals can be written in a highly visual "polynomial" form (see fig. 22. 2 for examples) that can reinforce students' understanding of "new" algebraic forms, such as $7x^2 + 3x + 2$. The following discussion demonstrates the relationships between algorithms in the two systems.

The Commutative and Associative Properties of Addition

Because the Aztec number system is nonpositional, the symbols for numbers that are larger than 10 and require two symbols, such as 15 (a lozenge and a bar) or 25 (a flag and a bar), can be reversed and still represent the same number. Teachers can use this property of Aztec numbers to illustrate the commutative property of addition for the real numbers, as shown in figure 22.3.

Similarly, the construction and interpretation of Aztec numbers that are larger than 10 and require three symbols, such as 35 (one flag, one lozenge, and one bar), can illustrate the associative property of addition for the real numbers, as shown in figure 22.4. Parentheses have been introduced to set off the two numbers that should be added first before their sum is added to the third number.

The Commutative Property of Addition

For any real numbers *a* and *b*, *a* + *b* = *b* + *a*.

Examples

In the real numbers: 10 + 5 = 5 + 10

In Aztec numbers:

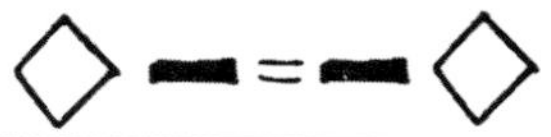

◇ ▬ = ▬ ◇

Fig. 22.3. The ordering of symbols in Aztec numbers illustrates commutativity.

The Associative Property of Addition

For any real numbers *a*, *b*, and *c*, *a* + (*b* +*c*) = (*a* + *b*) + *c*.

Examples

In the real numbers: 20 + (10 + 5) = (20 + 10) + 5

In Aztec numbers:

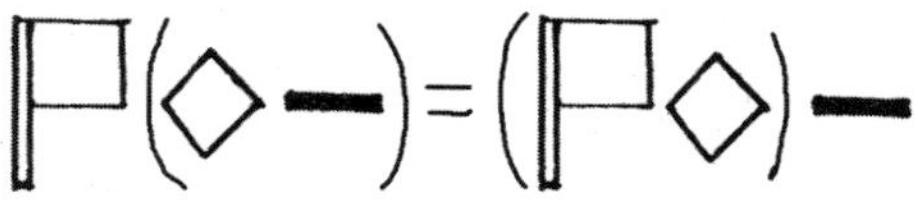

𝗣(◇▬) = (𝗣◇)▬

Fig. 22.4. The construction and interpretation of Aztec numbers illustrate associativity (parentheses for illustrative purposes only).

Polynomial addition

No available information shows how the Aztecs actually performed addition; however, Vasquez believes that addition of Aztec numerals involved combining like terms in a manner similar to that used in adding polynomials in algebra (for example, $[3x^2 + 5x + 7] + [5x^2 + x + 2]$). Figure 22.5 shows the addition of 37 and 45 in Aztec numerals. Symbols have been written in descending order of value from left to right, using a flag for 20, a lozenge for 10, a bar for 5, and dots for units. Parentheses have been used to group the quantities that are being added, and the addition sign has been introduced to denote the operation being performed.

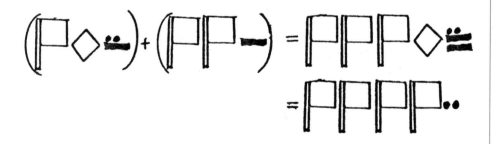

Fig. 22.5. Adding 37 and 45 in Aztec numbers by collecting like terms, as in adding polynomials in algebra

Polynomial multiplication

No historical evidence shows Aztec multiplication. Therefore, the method demonstrated in the example in figure 22.6 is only one possible way of carrying out that mathematical operation using Aztec numerals.

As the figure shows, the multiplication of 15 by 25 in Aztec numerals could be accomplished by using the FOIL method for the multiplication of two binomials—the familiar process that calls for multiplying *first, outer, inner,* and _last_ terms in sequence and then adding the products. This procedure can be generalized for multiplying any two Aztec numbers, no matter how many symbols they include. That is, the multiplication of Aztec numerals can proceed in exactly the same way as the multiplication of polynomials in algebra (for example, $[2x^2 + x + 1][x^2 + 3x + 2]$), requiring repeated applications of the distributive property and collecting like terms to simplify the result. (Teachers could illustrate algebraic multiplication of a polynomial by a monomial by showing the multiplication of an Aztec numeral expressed by two or more symbols, such as 15 [represented by a lozenge and a bar], by a numeral expressed by a single symbol, such as 5 [represented by a bar]. However, we omit that particular algebraic operation because it is embedded in the example shown in figure 22.6.)

The FOIL Method

In algebra, for any real number _a, b, c,_ and _d:_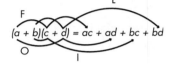

With real numbers: $(10 + 5)(20 + 5) = 200 + 50 + 100 + 25 = 375$
With Aztec numbers:

Fig. 22.6. Multiplying Aztec numbers using the FOIL method from algebra

---

Ideas for the Classroom

Teachers can assign to students the following questions either as classwork or homework. The purpose of the questions is to sharpen students' understanding of the associative, commutative, and distributive properties in algebra by applying them in an ethnomathematics context and to enhance students' understanding and interpretation of history. (Answers to these questions are also provided at the end of the article.)

1. How many years passed between the Aztecs' founding of their principal city, Tenochtitlan, and the beginning of the conquest of Mexico? Compute your answer using Aztec numerals, and write it at least two different ways by changing the order of the number symbols.

2. Using Aztec numerals, subtract 1200 from 1325. What is the significance of the years A.D. 1200 and A.D. 1325 in the history of the Aztecs? Provide an interpretation of the result of your subtraction in the context of Aztec history.

3. Add 292 to 1200 using Aztec numerals. Give an account of the 292 years that followed A.D. 1200 in Aztec history.

4. Multiply 25 by 30 using Aztec numerals, and collect like terms to simplify your result.

5. Multiply 15 by 35 using Aztec numerals, and collect like terms to simplify your result.

Teachers are encouraged to generate additional exercises that enable students to use the algebraic properties discussed in this lesson.

Recommendations to Teachers

The following recommendations are intended to assist teachers in presenting this or similar material in mathematics classrooms:

1. Teachers should choose the simplest symbols to represent numbers and limit their examples to numbers that involve no more than three symbols in the interest of minimizing the time and labor of writing large numbers with Aztec number symbols. Although teachers should generally adopt the convention of writing number symbols in descending order of value from left to right, rearrangements of the symbols can serve as starting points for the topics of combinations and permutations in discrete mathematics.

2. Before teachers discuss the number systems of different cultures and show maps of the regions where those cultures thrived, they should do some reading on ethnomathematics and the general history of the cultures so that they can integrate this information into their teaching of mathematics. This background reading will help teachers connect mathematics with history, geography, and social sciences while increasing their knowledge of ethnomathematics, world history, and cultures. Moreover, teachers may also use the opportunity to learn more about the cultures of students whose backgrounds are related to the cultures under discussion. Stuart (1981), Soustelle (1961), and Vaillant (1962) would be useful sources for teachers to consult for supplemental information on the Aztecs.

3. An additional use of the Aztec number system in elementary algebra classes is in assessment and evaluation activities. For instance, teachers can create exercises that are similar to the examples provided in this article for the purpose of assessing students' understanding of the algebraic properties illustrated here. Those questions can be either integrated into, or separate from, the tests or quizzes covering the corresponding course material in algebra. This practice will enhance a multicultural approach to the teaching of mathematics. Before formal assessments, teachers can design questions involving Aztec numerals to evaluate students' progress in learning the algebra skills and concepts discussed in this article.

4. Other number systems besides the Aztecs' can be used to teach precalculus concepts in a multicultural context. For instance, Egyptian numerals can be used to illustrate polynomial addition and multiplication, and the Egyptian and Aztec number systems can both be used to introduce combinations and permutations. Furthermore, the pre-Columbian Olmec number system can serve as a starting point in the teaching of exponential growth and exponential functions (see Ortiz-Franco 1993).

CONCLUSION

The inclusion of the number systems of other cultures in mathematics classrooms accomplishes several objectives in addition to enabling students to experience mathematics in an ethnomathematics context. Using Egyptian or pre-Columbian number systems in mathematics classes that include African American, American Indian, or Latino students shows these students that mathematics is an activity that has been undertaken throughout history by cultures other than Western civilizations. In addition, they will realize that mathematics is an integral part of their own cultural traditions. These observations may stimulate such students and motivate them to pursue their studies of mathematics, thereby ultimately increasing their representation in mathematics-based careers.

REFERENCES

Ascher, Marcia. *Ethnomathematics: A Multicultural View of Mathematical Ideas.* Pacific Grove, Calif.: Brooks/Cole Publishing Co., 1991.

Cajori, F. *The Early Mathematical Sciences in North and South America.* La Salle, Ill.: Open Court Publishing Co., 1928.

D'Ambrosio, Ubiratan. "Ethnomathematics: What Might It Be?" *International Study Group on Ethnomathematics Newsletter* 1, no. 1 (1985a): 2.

———. "Ethnomathematics and Its Place in the History and Pedagogy of Mathematics." *For the Learning of Mathematics* 5, no. 1 (1985b): 44–48.

Joseph, George Gheverghese. *The Crest of the Peacock: Non-European Roots of Mathematics.* New York: Penguin Books, 1991.

National Council of Teachers of Mathematics (NCTM). *Curriculum and Evaluation Standards for School Mathematics.* Reston, Va.: NCTM, 1989.

———. *Professional Standards for Teaching Mathematics.* Reston, Va.: NCTM, 1991.

National Science Foundation. *Women, Minorities, and Persons with Disabilities in Science and Engineering: 1996.* NSF 96-311. Washington, D.C.: National Science Foundation, September 1996.

Nelson, David. "Ten Key Areas of the Curriculum." In *Multicultural Mathematics: Teaching Mathematics from a Global Perspective*, edited by David Nelson, G. G. Joseph, and Julian Williams, pp. 42–84. New York: Oxford University Press, 1993.

Ortiz-Franco, Luis. "Chicanos Have Math in Their Blood: Pre-Columbian Mathematics." *Radical Teacher* 43 (fall 1993): 10–14.

Payne, Stanley E., and Michael P. Closs. "A Survey of Aztec Numbers and Their Uses." In *Native American Mathematics*, edited by Michael P. Closs, pp. 215–35. Austin, Tex.: University of Texas Press, 1993.

Soustelle, Jacques. *Daily Life of the Aztecs.* Stanford, Calif.: Stanford University Press, 1961.

Stuart, Gene S. *The Mighty Aztecs.* Washington, D.C.: National Geographic, 1981.

Vaillant, George C. *Aztecs of Mexico: Origin, Rise and Fall of the Aztec Nation.* Garden City, N.Y.: Doubleday & Co., 1962.

Answers to questions in lesson 1, p. 242:

1.

The year A.D. 1200 is the year when the Aztecs first appeared in the Valley of Mexico.

2.

In the year A.D. 1325, the Aztecs established their principal base in Tenochtitlan, now Mexico City.

3.

From the perspective of fifteenth-century Europeans living in the Eastern Hemisphere, the expedition that Columbus led, as the first such expedition from Europe, "discovered" the Western Hemisphere, which they did not know existed. To them, this was a "New World." However, considered from a global historical perspective, the notion that Columbus "discovered" the New World demeans the humanity and accomplishments of the people who had arrived in the Western Hemisphere thousands of years earlier. They were the ones who "discovered" it, and they created civilizations that thrived there for centuries before Columbus arrived.

4.

The year A.D. 1519 marked the start of the Spanish conquest of Mexico. By A.D. 1519, the Aztec empire ruled over millions of subjects and embraced lands from the Gulf of Mexico to the Pacific Ocean and from northern central Mexico to the present Guatemalan border.

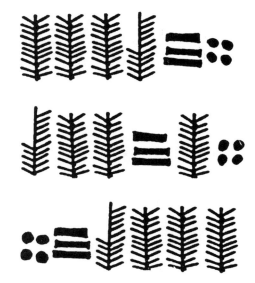

Answers to questions in lesson 2, pp. 244–245:

1.

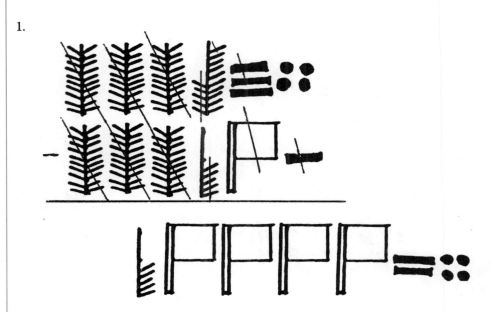

The Aztecs founded their principal city, Tenochtitlan, in A.D. 1325, and the conquest of Mexico began in A.D. 1519. Thus, 194 years passed between these two historical events.

2.

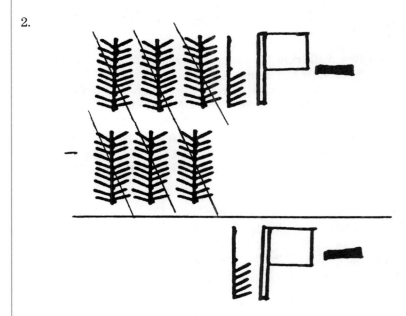

In the year A.D. 1200, that the Aztecs first appeared in the Valley of Mexico. In A.D. 1325, the Aztecs established their principal base, Tenochtitlan. Thus, the Aztecs took 125 years to establish their principal base from the time that they first arrived in the Valley of Mexico.

3.

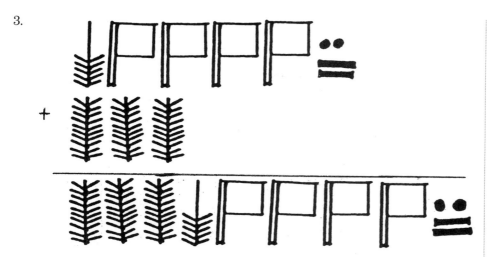

The 292-year period from A.D. 1200 to A.D. 1492 can be considered as encompassing the entire record of Aztec history. In this period, the Aztecs progressed from not having a permanent place to live to being, along with the Incas, one of the most influential civilizations in the New World.

4.

5.

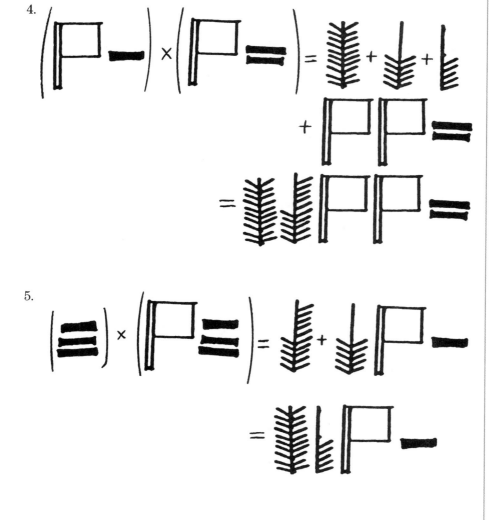

Computation, Complexity, and Coding in Native American Knowledge Systems

23

Ron Eglash

Research in the knowledge systems of indigenous societies has been hampered by both cultural and technological assumptions. We see these assumptions at work in many popular television documentaries that present the "vanishing native" who "lived at one with nature." Such portraits are well intended, but they serve only to reinforce the stereotype of indigenous peoples as historically isolated in a static past. The idea that these societies lived "close to nature" implies that they engaged in concrete rather than abstract thinking—that they were simple and "primitive," taking only the first steps up a supposed ladder of progress. We need to open our eyes to the dynamic histories and technological sophistication of indigenous cultures. For example, we need to think about the active, evolving ecological knowledge of indigenous people instead of focusing on the passive portraits drawn by statements such as "Indians lived as part of the ecosystem." Rather than cling to the illusion of a frozen precolonial tradition, we need to see indigenous societies as having always been in a state of change and to understand more recent features of Native American life as part of that history.

People often confuse biological and cultural evolution. Cultural evolution is the Lamarkian process by which we pass on from one generation to the next the knowledge that we acquire. By contrast, biological evolution is the Darwinian process by which the rare lucky mutant with an advantage then passes it on in DNA to the next generation. The time frames for these two processes are of different orders of magnitude. Significant biological evolution occurs very slowly, over millions of years, whereas dramatic cultural evolution unfolds in just a few thousand years. This difference explains why human beings have so little genetic variation: the first modern humans, from their single origin in Africa, quickly spread across the earth in a few thousand years. Our nearly identical genetic composition is a result of speedy Lamarkian cultural evolution's adapting us to these new environments.

Stephen Jay Gould (1981) explains that biological evolution used to be thought of as forming a single ladder of progress but is now seen as a "copiously branching bush" (p. 348). In the same way, cultural evolution is now typically portrayed as a branching diversity of forms. While European societies were following one particular sequence in the development of mathematics, other cultures were developing mathematical ideas along very different lines. Instead of assuming that Native American mathematics is restricted to simple counting systems or geometric forms, we should be open to other mathematical patterns that appear, including those that are embedded, emergent, or obscured by difficulties in translation to their western counterparts. Equally important, we should strive to show the interrelationships among such culturally embedded mathematical concepts. Through an examination of computation, complexity, and coding in Native American knowledge systems, this article attempts to show how such an approach can open new possibilities.

ALGORITHMIC COMPLEXITY AND BIODIVERSITY

Gary Nabhan, an agriculture researcher who works with Native American growers, notes that sustaining genetic diversity was an important theme in indigenous knowledge systems. In searching for crop variety, Nabhan found that indigenous people with the strongest focus on ceremonial religious practices were also those with the greatest diversity in genetic resources. For example, some groups propagated a rare variety of bean for a winter ceremony, sprouting it in underground kivas. A rare variety of sunflower had also been allowed to grow weedlike around the fields because its petals were used to make a yellow ceremonial face paint. Cutler (1944) found that South American medicine men conducted a ritual in which they were propagating "podcorn," a variety of corn that cannot grow wild because each kernel is covered by a heavy husk. Even outside these ceremonial settings, Native American farmers offered reasons that were rooted in religion for their efforts to promote genetic diversity (Nabhan 1983, p. 7):

> On one occasion, I asked a Hopi woman at Munqapi if she selected only the biggest corn kernels of all one color for planting her blue maize. She snapped back at me, "It is not a good habit to be too picky ... we have been given this corn—small seeds, fat seeds, misshapen seeds—all of them. It would show that we are not thankful for what we have received if we plant just certain ones and not others."

Why should Native American religions put such a strong emphasis on maintaining a complex variety of genetic resources? From a biological point of view, diversity turns out to be essential for coping with environmental uncertainty. The winter ceremony bean is very resistant to root-knot nematodes; this characteristic is critical in occasional years of nematode epidemics. The wild sunflower is not edible, but it can cross-fertilize with cultivated varieties and thus can increase the sunflower's genetic potential for adaptation. Cutler (1944) has suggested that Kokopelli, the humpbacked flute player of Southwestern iconography, is an herbalist from the South American tradition. History reports that these herbalists traveled north at least as far as Central America, and they still travel with flutes and backpacks today. Kokopelli is also regarded as a fertility figure, ensuring the reproduction of all living organisms. The link between plant diversity and animal diversity—a connection of great concern in today's environment—was given an important place in this traditional knowledge system.

This system's recognition of a necessary relationship between complexity and uncertainty was not restricted to plant and animal genetics. Native American creation myths exhibit a similar correspondence. The Navajo story of creation involves the creator/trickster, Coyote (Burland 1965, p. 93):

> First Man, First Woman, and Coyote ... were not satisfied with the sky.... So they searched for glittering stones and found some mica dust. First Man placed the Star Which Does Not Move [Polaris] at the top of the heavens.... Then he placed the four bright stars at the four quarters of the sky.... Then in a hurry, Coyote scattered the remaining mica dust so it did not fall into exact patterns but scattered the sky with irregular patterns of brilliance.

After people create order, Coyote creates randomness, tossing bits of rock into the sky. For Native Americans, "randomness" is not merely disorder: "random" tosses were precisely gauged in their gambling procedures. Ascher (1991) provides a vivid illustration of this fact in her description of the Native

American game called "dish." In the Cayuga version of the game, six peach stones blackened on one side are tossed, and the number landing black or brown side up is recorded as the outcome. The traditional Cayuga point scores for each outcome (rounded to whole numbers) are the exact values calculated by probability theory.

In the Navajo myth cited above, Coyote—acting in his usual haphazard way—creates rain and brings the seeds of all the plants. The idea of creating an "irregular," complex collection of genetic resources to match the irregular, complex randomness of natural events is thus deeply embedded in many Native American societies and in diverse aspects of their knowledge systems. It is also a foundational concept for certain measures of complexity used in modern mathematics. Examining the details of these complexity measures can shed light on their relationship to Native American knowledge systems.

The first mathematical models for complexity were developed in the work of Andrey N. Kolmogorov and Gregory J. Chaitin in the late 1950s (see Pagels [1988] for a popular historical review). Noting that some apparently random numbers can be completely determined by a simple algorithm, Kolmogorov and Chaitin proposed that the "algorithmic complexity" of a number was equal to the length of the shortest algorithm required to produce it. This statement means that periodic numbers (such as 0.121212121...) will always have a low algorithmic complexity. Even though the number is infinitely long, the algorithm can simply say, "Repeat 12 forever." A longer algorithm, like that required to produce ≠, would have a higher algorithmic complexity. Truly random numbers (e.g., strings of numbers produced by repeatedly rolling dice) have the highest algorithmic complexity possible, since their only algorithm is the number itself and since a number of infinite length gives infinite complexity.

From a mathematical point of view, the reason why Native American knowledge systems emphasize maintaining biodiversity is that algorithmic complexity increases with randomness. Thus, Coyote's random natural events can be matched only by the maximum complexity in genetic resources. But the parallels may be more than just analogy. Miguel Jiménez-Montaño (1984) devised what several researchers have regarded as one of the most workable systems for measuring algorithmic complexity—a system that he developed at the University of Veracruz in Mexico for use with amino acids and genetic sequences. Jiménez-Montaño credits his Ph.D. dissertation adviser, Werner Ebeling, as the principal scientific inspiration for his research. Yet Jiménez-Montaño was himself well aware of indigenous cultures' valuing of complexity: he was a longtime friend of Mario Vazquez, who is a distinguished botanist in the Biological Institute at the University of Veracruz and studies archaeological data on the plants used by the indigenous societies of Mexico. Native American complexity concepts may well have influenced Jiménez-Montaño's work, even if only subliminally or indirectly.

The difficulty in applying the Kolmogorov-Chaitin measure is that Kolmogorov and Chaitin used a class of universal symbol-generating systems called "Turing machines," named for British mathematician Alan Turing. Jiménez-Montaño's insight was in applying a more restricted set of procedures (what computation theorists call a *context-free grammar*) for generating the symbols (amino acid sequences), a method that makes it easier to ensure that one has actually found the algorithm of minimum length. In a context-free grammar, we begin with a starting symbol, S, and production rules that tell how to replace previous symbols with new ones. A list of "terminal symbols" tells

COMPLEXITY AND COMPUTATION: THE KOLMOGOROV-CHAITIN MEASURE

which symbols cannot be replaced. For example, given terminal symbols $\{a,c,t\}$ and rules $\{S \rightarrow cb, b \rightarrow ad, d \rightarrow t\}$, we get $\{cat\}$ as the only permissible string in this grammar. To measure the complexity (K), we just sum the number of symbols on the right-hand side of the production rules: $K = 2 + 2 + 1 = 5$. The complexity measure has just one more part: since a string of symbols that repeat should count less (recall what we said previously about the low complexity of periodic numbers), any production rule in which n repetitions of a single symbol occur is counted as $1 + \log_2 n$.

Jiménez-Montaño's work applied complexity measures to amino acid sequences, but we can just as easily apply them to biological phenomena in the everyday world. Corn is a particularly good example of such a phenomenon because one can easily see the contrast between the single variety of yellow corn that we eat and the diverse varieties of "Indian corn" that we use for decorative purposes in the fall. Geneticist Barbara McClintock discovered _transposition_—the release of a chromosome element and its insertion into a new chromosome—by observing the complex color patterns in Indian corn, and she later worked on National Science Foundation projects to preserve the Native American seed stocks that were threatened by the popularity of yellow corn (Keller 1983).

Actual corn patterns are typically quite irregular, but for the sake of our illustration, we will pretend that we have found some very regular rows that include only the colors yellow (Y), black (B), red (R), and white (W). Suppose that we compare two rows of 32 kernels each with the following repeating patterns:

1. YBRW ...

2. YBRWBWYB ...

Pattern 1 repeats every four symbols; pattern 2, every eight symbols. Intuitively, we see pattern 2 as more complex, and we can confirm our sense of this difference by using the complexity measure. The minimum set of production rules can be found by experimentation and are as follows:

For pattern 1:

$$S \rightarrow a^8 \text{ (meaning } aaaaaaaa),$$

$$a \rightarrow YBRW.$$

The complexity is $K = (1 + \log_2 8) + 4 = 7$.

For pattern 2:

$$S \rightarrow a^4,$$

$$a \rightarrow bc,$$

$$b \rightarrow YBRW,$$

$$c \rightarrow BWYB.$$

The complexity is $K = (1 + \log_2 4) + 2 + 4 + 4 = 12$.

This agrees with our intuition about the difference in the complexity of the two patterns. To learn more about the work of Jiménez-Montaño and his explorations of biological complexity, readers might examine his recent publications (e.g., Cocho et al. [1993]). Meanwhile, let's look again at the traditional Native Americans concepts. How might Jiménez-Montaño's approach be related to these indigenous knowledge systems?

COMPUTATION IN OJIBWAY SCROLLS

The Ojibway societies historically occupied a vast area extending from present-day Manitoba in Canada to Michigan and Minnesota in the United States. The Southern Ojibway created a pictographic method for recording their ideas—primarily those relating to cosmology and ritual narrative—by etching birchbark strips. Selwyn Dewney (1975), a researcher at the Glenbow-Alberta Institute in Calgary, Alberta, has published an extensive collection of these sacred scrolls and has remarked on particular numerological patterns in them. In a similar analysis, Closs (1986) has shown that several of the scrolls depict groupings in multiples of 4. Dewney also notes these patterns but focuses more attention on the multiples of 3 that appear in what he terms "deviant scrolls." Both authors correctly observe that multiples do occur, but the scroll patterns are in fact explained more fully by the kind of production rule generation system used by Jiménez-Montaño.

Figure 23.1 shows the diagram etched on a birch bark scroll by an Ojibway shaman named Sikassige. The diagram, originally published in Hoffman (1891), shows four stages of a shaman's initiation and "path of life" (Dewney 1975, p. 74). The depiction of each stage features a lodge presided over by several officials. Moving from right to left, we see that the number of officials at each lodge (the human figures on the outside edge of the rectangles) increases in the pattern 8, 12, 18, 24. Since 8 is not divisible by 3 and 18 is not divisible by 4, this sequence is not adequately explained by multiples. However, the sequence can be completely described as the result of production rules that are based on the groupings of the officials. Since the officials in each lodge are separated by spaces, we can think of them as symbol strings that generate new symbols according to the production rules below:

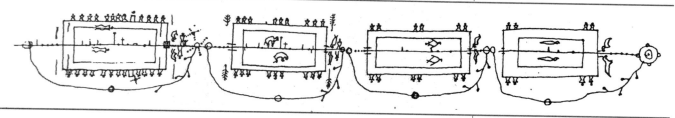

Fig. 23.1. Sikassige's sketch (Source: Dewdney 1975, p. 74, plate 54; used with permission of University of Toronto Press)

In the first lodge:

Each of the bottom and top pairs of officials yields one new official for the next lodge.

The total number of officials increases from 8 to 12.

In the second lodge:

Each of the bottom and top pairs of officials yields one new official for the next lodge.

The total number of officials increases from 12 to 18.

255

In the third lodge:

Each of the bottom and top triplets of officials yields one new official for the next lodge.

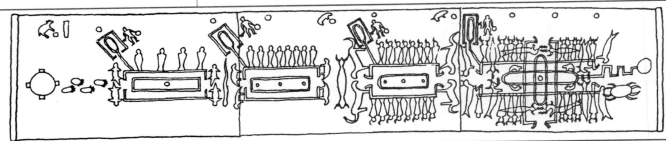

The total number of officials increases from 18 to 24.

Sikassige probably was not thinking of the sequence as consisting of self-generating symbol strings—for one thing, such thinking would require that the spaces between officials also be represented as symbols, a state of affairs that would make the production rules more arbitrary. The first rule would then have to place the space associated with right-hand pairs in mirror-reverse of the space associated with left-hand pairs (or else pairs on the right and pairs on the left would have to be replaced by two different rules). Nevertheless, Sikassige may well have thought of the sequence as self-generating, like the stages of initiation themselves, with each stage creating the preconditions for the next one.

Figure 23.2 shows another initiation or road-of-life scroll, which Dewney estimates was created between 1825 and 1875. Its increasing sequence of officials lends itself quite readily to description by production rules. Moving across the four lodges from left to right, we see that the number of officials (the human figures on the outside edge of the rectangles) increases at each lodge except for the last. Expressed as production rules, the configurations give us $b^n \rightarrow b^{2n}$ for the first iteration, $b^n \rightarrow s^{2n}$ (where s represents an official with a single horn) in the next iteration, and $s \rightarrow d$ (where d stands for an official with a double horn and acts as a terminal symbol). *Terminal symbol* is a term used by Dewney (p. 97) to discuss what Hoffman's informant, an Ojibway shaman named Skwekomik, called "end of the road" icons.

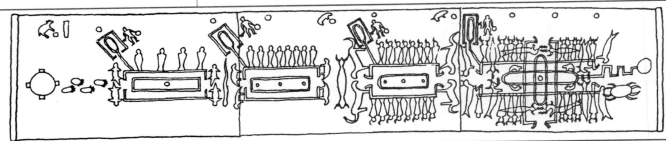

Fig. 23.2. Mide ceremony (Source: Dewdney 1975, p. 94, plate 77; used with permission of University of Toronto Press)

The Lac Court Oreilles scroll shown in figure 23.3 depicts ceremonial proceedings whose functions have not been fully identified. Interpretation requires close attention to the scroll's details. We see not only what we take to be the familiar officials increasing in number with each lodge but also some similarly shaped posts in the center of each lodge. Each post is either single or double,

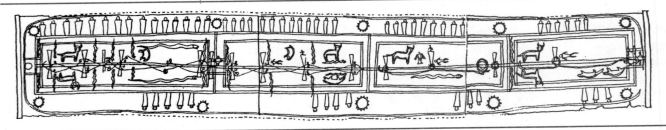

Fig. 23.3. Lac Court Oreilles Scroll (Source: Dewdney 1975, p. 134, plate 134; used with permission of University of Toronto Press)

and the posts in a particular lodge can be viewed as a specification of the production rules to be applied in that lodge. These rules are as follows:

In the first lodge:

The post in the center is single, so each of the three officials at the bottom generates one new official for the second lodge.

Thus, from the first lodge to the second, the total number of officials increases from 9 to 12.

In the second lodge:

Two posts in the center are double and one is single, so two of the three officials at the bottom generate two new officials apiece for the third lodge, and the third official at the bottom generates only one new official for the third lodge.

Thus, from the second lodge to the third, the total number of officials increases from 12 to 17.

In the third lodge:

The four posts in the center are double, so each of the four officials at the bottom should generate two new officials.

Thus, from the third lodge to the fourth, the total number of officials should increase from 17 to 25—but the scroll shows only 20.

The only exception to the production rules occurs in the last lodge. If our guesses about the rules are correct, what happened to the missing five officials? Dewney notes that in the final stages, initiates were warned that they might be diverted from the path by evil spirits, often appearing as serpents. Since five pairs of serpents oriented in the north-south (or "bad") direction appear in the inner rectangle, they may be all that is left of our five missing officials.

COMPLEXITY AND COMMUNICATION: THE SHANNON-WEAVER MEASURE

Just as genetic complexity can be measured using the mathematical theory of computation, entire ecosystems can be measured for complexity using the mathematical theory of communication. Such measures are increasingly important in keeping track of endangered environments (see Whittaker [1975]), and they also have parallels in both Native American biocultural practices and Native American mathematics. Ecologists first began measuring ecosystem complexity by counting the number of species per unit of area, but they found this method unreliable. Suppose there was an ecosystem in which there were 500 different species of animals but 99 percent of the individual animals were of one species? That would mean that if you took a trip there, your probability of seeing anything but members of the one common species would be quite low. The ecosystem would actually have low species diversity. One way to resolve this apparent paradox is to move it into the realm of communication. At an intuitive level, we can see that communicating about or describing a very simple environment is easier than communicating about or describing a complex environment. Looking for a way to state and verify this intuition, ecologists applied the mathematical theory of communication, proposed by Claude Shannon and Warren Weaver in the late 1940s, to develop a quantitative measure of ecological complexity.

Shannon and Weaver defined the basic unit of information transmitted in a communication system in terms of probability. Suppose you are lying in bed in the morning after a night for which the weather forecasts gave a 50 percent chance of snow. Suddenly, your mother calls out, "Get down to breakfast! Sun's out!" You would have just been provided with some useful information. Now suppose you went to bed in the middle of a blizzard, with the forecasts giving a 99 percent chance of snow for the coming day. The next morning, your little sister wakes you up to say, "Look, there's more snow falling today!" She would have provided you with less information, because telling something of which the hearer is already pretty sure is less informative than telling something about which the hearer is very uncertain. The less likely an event, the more information is communicated by its symbol. This relation is precisely defined by Shannon and Weaver: each symbol of probability (p) contributes an amount of information (I) in bits, $I = -\log_2 p$. To get the average number of bits per symbol, usually referred to as the entropy (H) for a given communication system, we need only take the sum of pI for each symbol. In the 1960s, ecologists noticed that H would make a good way to evaluate the diversity of an ecosystem, because it combines the contribution that a species makes to diversity with its total population. That is, H reflects not only how many species there are but also how much of each species there is. "How much" can be measured in many different ways besides the number of individuals per unit of area and is referred to as *species importance* (see Whittaker [1975]).

Suppose, for example, that we have two ecosystems in which we have three birds: sparrows, robins, and crows. We can think of the probabilities as the population percents (expressed as decimal fractions) for each:

1. $S = .10, R = .15, C = .75$

2. $S = .33, R = .33, C = .34$

$$H(1) = (.10 \times 3.322) + (.15 \times 2.737) + (.75 \times 0.415) = 1.054 \text{ bits}$$

$$H(2) = (.33 \times 1.600) + (.33 \times 1.600) + (.34 \times 1.556) = 1.585 \text{ bits}$$

These calculations support our intuition that ecosystem 1, overwhelmed with crows, is not as diverse as ecosystem 2. The maximum amount of information per symbol is obtained when all symbols are equally probable. More generally, we can see the relationship between complexity and communication: the more complex the system, the greater the amount of information is needed to convey its description. This requirement leads quite easily to the concept of optimal coding, because you cannot use communication to gauge complexity if some of the description is superfluous. Many Native American aphorisms regarding communication refer to this concept of optimal coding, such as, "It does not require many words to speak the truth" (Chief Joseph, Nez Percé). Native American habits of speech are often contrasted with Euro-American tendencies to be chatty or verbose (see Basso [1979]).

Teachers interested in using Native American knowledge systems in mathematics classrooms will find a rich resource in applications of information theory to indigenous coding practices. Calculations of bits per message (*entropy*) or bits per second (*channel capacity*), for example, can be carried out for sign language, smoke signals, bead and feather patterns, sand paintings, and other media of Native communication (see Witherspoon and Peterson [1995] and Mallery [1972] for Native American coding examples and Pierce [1980] for an introduction to information theory). As a quick illustration, let's compare smoke signals with fire arrows. Mallery (1972) reports a wide variety of smoke-signal patterns, including some resembling "the telegraphic alphabet" (p. 537). Most appear to be based on the number of columns of smoke (that is, simultaneous fires) and the length of the columns. Fire arrows, a similar signal system used only during the night, can also appear simultaneously in groups and can be further distinguished by their vertical or diagonal orientations. Which can convey more code signals: a smoke system using a maximum of two columns of smoke (disregarding order) and one of three possible column lengths for each fire or a fire-arrow system using a maximum of three arrows and two possible orientations? This is a problem in combinatorics; here are all nine possible combinations:

- Smoke signals with long (L), medium (M), and short (S) columns:
  LL, LM, LS, MM, MS, SS, L, M, S
  Nine total code signals

- Fire arrows with vertical (V) or diagonal (D):
  VVV, VVD, VDD, DDD, VV, VD, DD, V, D
  Nine total code signals

Given equal probability for all $n = 9$ signals, $H = \log_2 n = 3.17$.

Now let us consider the Apache system (Mallery 1972, pp. 538–39), in which signals fall into three categories of warning: "attention" (no real warning), "safety" (mild warning), and "caution" (strong warning). Within each category, it appears that distinctions can be made on the basis of smoke-column length, whereas the categories themselves are distinguished by the number of simultaneous fires (from one to three). Variation in column length is not entirely clear, but let us assume that there are two equally probable possibilities, intermittent or continuous, and that attention signals are sent 55 percent of the time, safety signals are sent 30 percent of the time, and caution signals are sent 15 percent of the time. We can then calculate:

$$H = -\left(2\left(\left(\frac{.55}{2} \right) \log_2 \frac{.55}{2} \right) + 2\left(\left(\frac{.30}{2} \right) \log_2 \frac{.30}{2} \right) + 2\left(\left(\frac{.15}{2} \right) \log_2 \frac{.15}{2} \right) \right) = 2.41$$

If lighting two fires takes twice as long as one, and lighting three fires takes three times as long as one, what is the optimal assignment of number of fires to categories? Our intuition correctly tells us that optimal coding (maximum rate of information transmission) would require that the frequency of use of a symbol should be inversely proportional to the speed of its signal, but let us confirm this supposition mathematically. The information rate R is defined as the average number of bits per second for a given communication system, and it is obtained by dividing H by the average number of signals per second. Using f for the number of seconds needed to send the signal (create a fire and generate the smoke), we have six possibilities for the information rate:

$$R_1 = \frac{2.406}{.55f + .30*2f + .15*3f} = \frac{1.504}{f}$$

$$R_2 = \frac{2.406}{.55*f + .30*3f + .15*2f} = \frac{1.375}{f}$$

$$R_3 = \frac{2.406}{.55*2f + .30f + .15*3f} = \frac{1.301}{f}$$

$$R_4 = \frac{2.406}{.55*2f + .30*3f + .15*f} = \frac{1.119}{f}$$

$$R_5 = \frac{2.406}{.55*3f + .30f + .15*3f} = \frac{1.070}{f}$$

$$R_5 = \frac{2.406}{.55*3f + .30*2f + .15*f} = \frac{1.003}{f}$$

Code R_1 is optimal and represents the code assignments recorded for the Apache system: "attention" signals use one fire; "safety" signals, two; and "caution" signals, three. Apache reasoning for this choice may have combined the idea of the frequency of use with some concept of the urgency of the message (the more fires, the more important the information) and perhaps even a notion of reducing error from false signals (three fires being the least likely to happen by accident). All these criteria would involve the concept of optimal coding.

NATIVE AMERICAN COMMUNICATION AND THE HISTORY OF COMPUTING

Looking at Native American complexity concepts, we find ties to contemporary computation in the work of Jiménez-Montaño. Are connections evident through communication, as well? Modern computing began with a synthesis between the development of production-rule generation theory and the engineering of machines capable of performing the required symbol manipulations. John von Neumann, who created the first synthesis, credited Alan Turing with having contributed the requisite mathematical theory of computation (see Hodges [1983, p. 304]). Turing was not only a theorist, however; in World War II, the British government required him to work on cryptography—specifically, on cracking the codes produced by the German army's "enigma machine." Meanwhile, the German cryptographers were stymied by an American military information system supplied by Native American "code talkers."

This period in American history did not represent the first time that Native Americans had been asked to provide U.S. military coding. Choctaw men had relayed messages during World War I through field telephones in France. In 1940, the Army Signal Corps ran tests with Comanches from Michigan and Wisconsin, and subsequently, Choctaw, Kiowa, Winnebago, Creek, and Seminole soldiers employed native languages to encrypt radio communications in Europe and North Africa. The most famous Native American participants in the war were the Navajo code talkers (Kawano 1990). Working together with army cryptographers, they developed a complex system that included both alphabetic symbols and encrypted whole words. The creation of the Navajo codes, as well as the training techniques for using them, combined indigenous knowledge systems with modern communication theory.

To state simply and without qualification that Native American code talkers led to the first computer would be absurd. Any Native American influence on the cryptography that occupied Turing during the war—and even Turning's influence on von Neumann—is diffuse, indirect, and subtle. But simply to write

Native Americans out of this history would be equally absurd. The process of invention is always carried forward on the crosscurrents of many different intellectual streams, and cryptography—with all its confluence of cultures—played an important part in the invention of computers.

The end of World War II did not terminate the connection between Native American culture and advancing computer technology in the United States. Defense systems engineer Tom Ryan married into a Navajo family living on a reservation in Arizona. He and his father, John Ryan, a Lockheed senior scientist, began reflecting on the wartime coding effort. If this kind of involvement was possible in wartime, why not in peacetime, as well? They launched the Navajo Technologies Corporation, which combined training at the Navajo Community School in Birdsprings, Arizona, with research on Ada compilers funded through several Department of Defense contracts (Able 1988).

This link was just one among a growing number of connections between Native American communities and the expanding computing field. In Alaska, for example, environmental conditions promoted the early use of distance learning, and the advent of home computers has resulted in the emergence of a widespread virtual community of indigenous people. The American Indian Computer Art Project (www.aicap.org/aicap/niiwin.html) features the work of indigenous artists who design in digital media, and the site even sells a digitizing stylus that is wrapped in traditional bead patterns. A more general Web site, NativeTech (www.nativetech.org), is "dedicated to disconnecting the term 'primitive' from perceptions of Native American technology and art."

Tribal Web sites have blossomed across the Internet, and some are working on technical projects in areas ranging from linguistics to ethnomathematics. Native Seeds/SEARCH (www.nativeseeds.org), for example, is a botanical organization dedicated to perpetuating the indigenous plant stock, and it has been creating a "cultural memory bank" that will tie Native American farmers to the agricultural contributions of their ancestors. The project, originated by Philippine ethnobotanist Virginia Nazarea (1996), documents a combination of cultural and biological information about the crops and their seeds, farming, and use. The information, including video interviews, is stored on CD-ROM, with access controlled entirely by the indigenous farmers. Describing this development brings us full circle, with the modern computer used to help maintain the biogenetic complexity fostered by Native cultures.

CONCLUSION

The mathematics portion of ethnomathematics has witnessed an extraordinary proliferation of creative yet rigorous frameworks. An exploration of mathematical activity has generated many new teaching resources. Yet the cultural dimension of ethnomathematics has received far less attention. In particular, portraits of Native American "tradition" often represent a static, homogenous society, lost in the distant past, despite the fact that critiques of such frameworks have been a focus of anthropology for at least a decade (see Clifford [1988]). Like these critiques, this essay attempts to broaden the view of Native American ethnomathematics so that we can, when appropriate, see "change" as traditional, "authenticity" as part of colonial politics, and the artificial worlds of mathematical technologies as extensions of sacred space.

REFERENCES

Able, Dawn. "The Navajos: Using Language for Two Nations." *Defense Computing* (May-June 1988): 35–38.

Ascher, Marcia. *Ethnomathematics: A Multicultural View of Mathematical Ideas.* Pacific Grove, Calif.: Brooks/Cole Publishing Co., 1991.

Basso, Keith. *Portraits of "The Whiteman": Linguistic Play and Cultural Symbols among the Western Apache.* New York: Cambridge University Press, 1979.

Burland, Cottie. *North American Indian Mythology.* London: Paul Hamlyn, 1965.

Clifford, James. *The Predicament of Culture.* Cambridge, Mass.: Harvard University Press, 1988.

Closs, Michael P. "Tallies and the Ritual Use of Number in Ojibway Pictography." In *Native American Mathematics*, edited by Michael P. Closs, pp. 181–212. Austin, Tex.: University of Texas Press, 1986.

Cocho G., F. Lara-Ochoa, Miguel A. Jiménez-Montaño, and J. L. Ruis. "Structural Patterns in Macromolecules." In *Thinking about Biology: An Introductory Essay*, edited by Wilfred D. Stein and Francisco J. Varela. New York: Longman, 1993.

Cutler, Hugh. "Medicine Men and the Preservation of a Relic Gene in Maize." *Journal of Heredity* 35 (1944): 291–94.

Dewdney, Selwyn. *The Sacred Scrolls of the Southern Ojibway.* Toronto: University of Toronto Press, 1975.

Gould, Stephen Jay. *The Mismeasure of Man.* New York: W. W. Norton & Co., 1981.

Hodges, Andrew. *Alan Turing: The Enigma.* London: Burnett Books, 1983.

Hoffman, W. J. *Middéwein or Grand Medicine Society of the Ojibway.* Seventh report of the U.S. Bureau of Ethnology to the Smithsonian Institute. Washington, D.C.: n.p., 1891.

Jiménez-Montaño, Miguel A. "On the Syntactic Structure of Protein Sequences and the Concept of Grammar Complexity." *Bulletin of Mathematical Biology* 46, no. 4 (1984): 641–59.

Kawano, Kenji. *Warriors: Navajo Code Talkers.* Flagstaff, Ariz.: Northland Publishing, 1990.

Keller, Evelyn F. *A Feeling for the Organism.* New York: W. H. Freeman & Co., 1983.

Mallery, Garrick. *Sign Language among North American Indians.* The Hague: Mouton, 1972.

Nabhan, Gary. "Kokopelli: The Humpbacked Flute Player." *Coevolution Quarterly* 37 (September 1983): 4–11.

Nazarea, Virginia. "Fields of Memories as Everyday Resistance." *Cultural Survival Quarterly* 20 (spring 1996): 61–66.

Pagels, Heinz R. *The Dreams of Reason: The Computer and the Rise of the Sciences of Complexity.* New York: Simon & Schuster, 1988.

Pierce, John. *An Introduction to Information Theory.* New York: Dover, 1980.

Whittaker, Robert H. *Communities and Ecosystems.* New York: Macmillan, 1975.

Witherspoon, Gary, and Glen Peterson. *Dynamic Symmetry and Holistic Asymmetry in Navajo and Western Art and Cosmology.* Bern, Switzerland, and New York: Peter Lang Publishing, 1995.

Yup'ik Culture and Everyday Experience as a Base for School Mathematics

24

Jerry Lipka

From 1989 to 1999, the author of this article and the authors of the following one, Esther A. Ilutsik and Claudia Zaslavsky, collaborated with Yup'ik elders to develop an ethnomathematics curriculum based on the indigenous Yup'ik culture and everyday experience. Authors Lipka and Ilutsik were, respectively, an "outsider" and an "insider" in this culture. Together with Zaslavsky, a specialist in ethnomathematics, we joined Yup'ik elders in situ as they practiced subsistence activities. "Subsistence" in this context describes a way of life that is related to hunting, fishing, and gathering. We observed and discussed these practices with the elders before beginning the complex, ongoing task of adapting them to a classroom context.

The purpose of our joint work has been to develop ethnomathematical activities for the elementary school classroom, prepare teachers to use elders' knowledge, and incorporate the local culture, language, and knowledge into classroom settings. We have been assisted in our work by a small but growing cadre of Yup'ik teachers from the Bristol Bay area of southwest Alaska who compose a research group called Ciulistet, which in Yup'ik means *leaders*. Through ethnomathematics—the study of the mathematics associated with different cultural groups—we have been responding to questions and calls for reform of standard schooling such as those voiced by Gerdes (1997, p. 225):

> How can this "totally inappropriate education, leading to misunderstanding and sociocultural and psychological alienation," be avoided? How can this "pushing aside" and "wiping out" of *spontaneous natural, informal, indigenous, folk, implicit, non-standard,* and/or *hidden (ethno)mathematics* be avoided?

Cultural groups have their own ways of perceiving the world and grouping, measuring, and categorizing its phenomena. As Bruner (1996) states, "learning and thinking are always *situated* in a cultural setting and always dependent upon the utilization of cultural resources ... in cultural communities [to] create and transform meanings" (p. 4). The notion of culture, cultural communities, and shared meaning guides what we write in these articles. We are fully aware that we have far to go to understand the present-day Yup'ik cultural context and how to connect what we learned to what students learn—and need to learn—in school. In presenting our approach as an alternative to education that isolates local knowledge and makes it alien, we acknowledge the work of many Yup'ik elders who have generously given us their time so that the next generation will *know*.

DEVELOPING CURRICULA FOR AND WITH INDIGENOUS PEOPLE

The growing interest in ethnomathematics related to the indigenous peoples of America is evidenced by this publication and others (e.g., Bazin and Tamez [1995]; Bradley [1984], [1993]; Caduto and Bruhac [1988]; Denny [1986]; Lipka [1994]; Pinxten, van Dooren, and Harvey [1983]; Zaslavsky [1996]). This expanding attention is encouraging to those of us who are committed to developing curricula and pedagogy for and with American Indians in general and Alaska Native people in particular. Simultaneously, however, the increased interest signals a need for caution and critical examination of the approaches taken and their potential impacts. Although research and curricula related to indigenous knowledge have become more common, indigenous people are not always the beneficiaries of this work. In fact, although the extant indigenous peoples of America, to varying degrees, retain their culture, including their languages and everyday practices related to living from the land, the majority of the new materials, curricula, and pedagogical approaches treat these groups as if they existed only in the past.

DILEMMAS AND CONTRADICTIONS

Even when new curricula are developed for and with indigenous people, a number of dilemmas and contradictions persist. Foremost among the problems is the fact that traditional knowledge is sometimes so scattered that cultural transmission from one generation to the next is no longer assured (Meade 1990). Present-day cultural communities are rapidly evolving as contemporary life embraces both subsistence and postmodern customs. Moreover, even when traditional knowledge is known, the process of interpreting it so that it fits school settings and connects with Western knowledge can result in cultural distortions (Lipka 1994; Pinxten 1997). Pinxten's account of mathematics education unfolding in a Navajo setting points to both cultural distortions and cultural alienation. Pinxten describes

> the utter ineffectualness [and] the total inappropriateness of the same material [Western curriculum fitting a Navajo context].... [Furthermore,] teachers I talked to showed a very poor understanding of the difference in outlook, system, preconceptions, and actual notions between the so-called Western (e.g., Piagetian) and the Navajo way. (P. 374)

Pinxten succinctly states the problem confronting non-Western education in general and Navajo education in particular:

> The impact and precise content of preschool native knowledge is different in the Navajo culture than it is in ours, and consequently the curriculum and its actual implementation in the Navajo situation should be adjusted in order to reach a smooth, understandable, really integrated development in the knowledge acquired at school. (P. 375)

Even under the best circumstances, it is extremely difficult to take an oral tradition associated with a subsistence lifestyle and transform it to fit the model of Western schooling. Seasonality, place, time, and culture organize activity within a subsistence lifestyle; in Western schooling, formal written literacy has precedence over oral expression, and a sequential, decontextualized curriculum organizes classroom activity. The task of developing a culturally authentic ethnomathematics curriculum faces many pitfalls. In school settings, indigenous knowledge, its context, and the language used to transmit it may all differ from their original counterparts. Concerns about such matters are not merely idle theoretical speculations. Precious limited resources are currently being expended on such curricular approaches as direct translation, which gives English "equivalents" for Yup'ik words without regard for cultural differences, cultural meanings, and the importance of context in supplying meaning. Another common approach involves simply replacing one term in a mathematics word problem by another assumed to be "culturally relevant"—for example, replacing *apples* with *salmon*. Recognizing the possibilities and limitations of such approaches is one of the challenges facing curriculum developers.

Despite these considerable dilemmas, using a collaborative ethnographic approach to developing ethnomathematics curricula and pedagogy has the potential to reinforce traditional cultural practices and promote pedagogy and curricula that are rooted in a Native American or Alaska Native cultural and cognitive framework. To avoid the negative consequences of *transforming* an indigenous cultural framework, we propose *grounding* the development of an ethnomathematics curricula in the present-day enculturation process. Developing curricula in situ, through activities and dialogue with elders, militates against trivializing these activities. "Situated" activities and dialogue have enabled us to establish methodological and curricular guideposts. A crucial assumption supporting these guideposts is the notion, gained from working with elders, that education should prepare students to live successfully both at home and in the larger world. One of the elders in the Ciulistet group, Henry Alakayak, often explains that individuals who can speak both Yup'ik and English are in a better position to understand the world and thrive in it. This "both/and" (see Lipka and McCarty [1994]) approach to curriculum calls on teachers, students, and curriculum developers to understand cultural systems and to apply that knowledge in an orderly way. This approach differs sharply from an "either/or" approach, which typically neglects local knowledge and practices.

A "both/and" approach requires an understanding of Yup'ik knowledge, language, and practice as they relate to ethnomathematics. Such an approach extends the tradition of Gay and Cole (1967), Pinxten (1997), and Gerdes (1997). Thirty years after the work of Gay and Cole, Gerdes (1997) quoted them on their conviction that it is

> "necessary to investigate first the 'indigenous mathematics,' in order to be able to build *effective bridges* from this indigenous mathematics to the new mathematics to be introduced in school: ... the teacher should begin with materials of the indigenous culture, leading the child to use them in creative ways." (P. 225)

Besides building "effective bridges" to the "new mathematics" associated with formal schooling, work needs to continue to explicate and develop an indigenous knowledge base. These goals can be accomplished through collaborative work with elders, indigenous teachers, and university consultants. Grounded research and practice (engaging with elders as they conduct their seasonal activities) can enable us to accomplish four objectives:

1. To learn the intuitive and embedded mathematics that exists within everyday cultural practice in context
2. To infer the cultural rules related to mathematical processes, concepts, and categories
3. To integrate indigenous context into curriculum that grounds classroom activity in, and connects it with, local community practice
4. To develop an interpretive and creative approach that works from an indigenous context and knowledge base and moves beyond the culturally familiar to "standard" Western mathematics while retaining some connections with the indigenous context and culture.

CURRICULAR OBJECTIVES OF GROUNDED RESEARCH

Second-order ethnomathematical classroom practice may involve students in making mathematical tools or following mathematical procedures that originate in some aspect of local practice or thinking but have been abstracted from it in such a way that they no longer retain an intimate connection to the culture and its context. In fact, such second-order practice may set a local context or foundation for mathematical tools that are not themselves used within the local culture. For example, teachers might invite students to make an abacus in such a way that it reflects Yup'ik numerical conventions and linguistic rules, even though an abacus is not itself a traditional Yup'ik instrument. Another activity might encourage students to develop pattern blocks that use the local culture's patterns and symbols but abstract them from their traditional use or otherwise decontextualize them. The links that these activities retain to the local culture allow the activities to serve as bridges back and forth between indigenous practices and Western mathematics. In effect, these mathematical activities are "backward compatible"—that is, they relate *back* to the local culture even though they are not *of* it. Again, such an approach follows some suggestions of Gerdes (1997) and Gay and Cole (1967), and the examples cited here will be discussed in more detail later.

Classroom practice developed from within an indigenous cultural and linguistic perspective can also provide students in the larger society with fresh insights and ways of perceiving mathematical relationships and solving problems. Our long-term collaborative research with Yup'ik elders, teachers, and school districts gradually evolved sufficiently to allow us to begin developing mathematics curriculum based on Yup'ik culture. This article describes our approach, which resolves or reduces some of the dilemmas previously outlined. By understanding how mathematics is embedded in indigenous practices and how concepts are shaped by indigenous language, we can construct authentic ethnomathematics for the classroom. The article gives detailed examples from our work with Yup'ik elders to highlight the process of our collaborative action research and the method and products that it has provided for an authentic, culturally mediated curriculum and pedagogy (Hollins 1996). We believe that the lessons learned from our work can be applied to teacher preparation (preservice and in-service) and to the development of classroom practices that reinforce both traditional Alaska Native culture and the teaching and learning of mathematics.

In addition, the article stresses the importance of having students who are members of the indigenous culture produce implements and take part in activities that have mathematical knowledge and relationships embedded in them and that are related to their culture. The article is organized according to the four objectives of grounded ethnomathematics research and practice enumerated earlier, and it shows how we have been pursuing these goals in the process of developing curriculum and pedagogy for in-service teachers.

BACKGROUND

The Bristol Bay area, where the work reported in this chapter has been taking place, is in southwest Alaska, approximately 350 miles southwest of Anchorage. The area is not well connected by road, and air travel is the major means of transportation to and from major population centers. The Yup'ik language is still spoken by a majority of the local adults, including teachers of Yup'ik background—a group that is steadily growing. However, English is the primary language spoken by the majority of the area's children. Most children and adults continue to be involved in seasonal subsistence activities, as well as activities that are linked to postmodern society and use modern technology.

In 1987, the author and the Ciulistet group began methodically exploring the mathematics related to the Yup'ik culture, language, and everyday practice. They called on Yup'ik elders to demonstrate traditional ways of measuring involved in making clothing, housing, and kayaks (Lipka 1994). The elders continued to teach the group, explaining indigenous methods and cultural meanings of enumeration and pattern making. This work is part of larger curricular and pedagogical efforts by the Ciulistet, the University of Alaska Fairbanks, and, more recently, two projects funded by the National Science Foundation: Alaska's Rural Systemic Initiative and an instructional materials development project in mathematics for elementary schoolchildren.

Because the primary activities of Alaska Native society remain subsistence-oriented, coordinated with seasonal fluctuations and associated with hunting, gathering, and fishing, they provide excellent opportunities for learning the skills and knowledge associated with subsistence and for becoming acquainted with the mathematics embedded in everyday tasks. Elders are particularly well versed in the art of fashioning implements for use in seasonal subsistence activities. Our exposure to this knowledge, which is intimately connected to the land and the seasonal cycles, has been essential to our development of an ethnomathematics curriculum. We have become aware of the elders' wealth of knowledge (see Fienup-Riordan [1994]; Kawagley [1995]; Lipka [1998]), which has opened up rich possibilities for curriculum and pedagogical development. This section shows some of our understandings about the mathematics associated with indigenous practice and language and how we have used this knowledge to form a conceptual basis for developing curriculum.

Mathematics Embedded in Cultural Practice: Building Blocks for an Ethnomathematics Curriculum

In our observations and work with elders, we have learned that living from the land requires Yup'ik people to make their own traps, snares, and other tools according to individual standards. Elders spend considerable time preparing tools for hunting, making clothing, and building such things as fish racks. In the process of designing and fashioning these implements, Alaska Native ways of knowing and being intersect most clearly with mathematical thinking. For example, Alaska Native people create standard measures that are quite distinct from such common linear measures as inches or centimeters. These indigenous measures attempt to accommodate both commonality (society) and individuality. An interest in both the individual and the group marks a culture that teaches how to measure and shape in general (that is, the socially accepted methods for measuring) at the same time that it teaches how to individualize these measures to suit a particular user and use. Yup'ik measuring calls for planning, visualizing, estimating, and using proportionality and ratios, all of which activities occur before end use.

In contrast with Alaska Native people, most of us today are exclusively consumers. We do not fashion our own tools or clothing but are only end users. Similarly, teachers and schoolchildren are mostly end users of consumables as they address themselves to mathematical tasks. For example, they use premade pattern blocks or other manipulatives. However, in teaching our group, the Alaska Native elders almost always acted as producers. They customarily fashioned uneven natural materials into implements and clothing through processes that required them to visualize end uses, shapes, designs, and space. We think that this "productive" phase, typically associated with seasonal subsistence

THEORY FROM PRACTICE: ACTION RESEARCH

activities, ought to be included in a program of ethnomathematics education. The following examples are taken from a course developed for the Ciulistet group during 1992 and 1993 at the University of Alaska Fairbanks.

Ethnographic Examples: Yup'ik-Based Mathematics

A Yup'ik elder, Lily Gamechuk, asked a teacher in our group to stand up so that the elder could measure her for a *qaspeq* (Yup'ik dress). After a very brief time, during which the elder never touched the teacher or used any instrument to measure her, the teacher was asked to sit down. Having used only her powers of her observation, the elder immediately cut out paper "clothing" that fit the teacher. (Since this was a spontaneous demonstration, authentic materials were not used.) The elder used her knowledge of relationships among body parts, both symmetrical and asymmetrical, and created a cognitive set of measures (using her own body as a standard) that she could add to or subtract from according to the measured person's dimensions.

Another time in our work with the Ciulistet, a group of male elders constructed a wooden fish rack, used for drying fish in the summer. Figure 24.1 shows such a rack. In this instance, we chopped down a few trees and cut them into workable lengths. From these pieces we determined a standard that became the measure against which we measured each subsequent piece of wood. This process ensured that all parts of our fish rack were in proportion to one another. In addition, the height of the rack was measured against the height of the end user, so that the fish rack would fit that person. Again, attention to body proportions, created standards, end tasks, and environmental conditioning were all part of the process of measuring.

The making of the dress and the building of the fish rack illustrate the Yup'ik interest in accommodating both individuality (by fitting the product or implement to the person) and society as a communal whole (by applying shared methods for measuring). Engaging students in building models of (or actual) fish racks and smokehouses are ways of involving them in cultural practices that are rich in embedded mathematics. Reflective inquiry can make the students conscious of the embedded mathematics. This awareness is what we are attempting to foster in the curriculum modules that we are developing for the National Science Foundation.

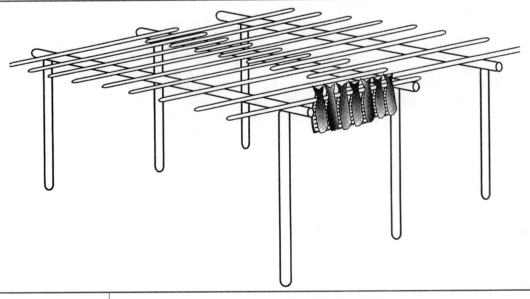

Fig. 24.1. Open fish rack

Another example underscores the role of body measures and measurement standards in construction tasks. In this instance, a Yup'ik elder showed us how a few parts of a kayak are customarily fashioned. He showed a body measure equal to the width of a kayak and explained that each person's kayak would be a different length depending on the individual's body measures but that everyone would use the same standard measure to determine the width of the kayak.

In each situation, the elders used their hands for both "static" and "dynamic" measuring. For example, the Yup'ik custom of using four fingers to equal the width of a person's foot allowed the elders to accomplish what can be thought of as static measuring. By contrast, when they measured by visualizing, which included actions such as working their hands into various shapes and constantly measuring the raw material being fashioned into finished products, they engaged in a far more dynamic process. Thus, measuring in Yup'ik culture is both static and dynamic, and it uses complex and multidimensional sensory data as the toolmaker coordinates his or her actions and motions to visualize— and thereby evaluate—such factors as volume, width, smoothness, unevenness, and roundness in relation to the end user and the end task.

Developmental Dimensions

Providing students with a measuring experience of this type is a bit like implementing Montessori's (see Gettman [1987]) developmental and sensorial approach to early mathematics—particularly geometry—with young students. However, a Yup'ik measuring experience teaches students that measuring is a series of multidimensional and complex acts instead of a static, single act, such as putting a square in its proper space, as in the Montessori approach. By involving students with adults and supporting them through peer-assisted instruction, teachers can enable students to become producers, fashioning their own tools.

These simple examples are strongly suggestive of a developmental approach to mathematics. For instance, after copying elders as they create border patterns for their clothing, students can fashion these patterns into classroom manipulatives. By creating the patterns and manipulating them, students can begin to build an intuitive understanding of geometric shapes and relationships. As producers as well as end users, students will learn more about the properties of raw materials, their limits, and their potential uses as they design the shapes needed for a particular task, such as making a border for traditional Yup'ik clothing.

Classroom activities that involve Yup'ik students in making such patterns according to traditional methods (creating a standard and making each piece relate to every other one) can help students learn about their culture and its embedded mathematics at the same time that they are gaining a feel for geometric relationships. This intuitive grasp can serve as an introduction to formal instruction in geometry at a later date. Visualizing and learning about geometry and spatial relations are some of the topics that researchers report as having distinct promise for advancing the mathematical education of Alaska Native or American Indian students (Pinxten 1997; Scollon 1979). In fact, MacPherson (1987), a teacher in the Canadian Arctic, reports on one student's "extraordinary ability to deal with shape, with space, and with size, which seemed inconsistent with his general disinterest in the number-based mathematics we offered in school" (pp. 24–26).

Increasing students' interest in mathematics by incorporating indigenous activities is exactly what we have been attempting to achieve in the curriculum we have been working on for the National Science Foundation's Division of Instructional Materials Development. In a classroom context, teachers and elders can introduce mathematics modules by demonstrating how they produce specific tools. In the example of patterns for clothing, an elder can show students how she works with strips of fabric to form a base piece (often a square or rectangular shape) from which she then creates the other pieces for her pattern. Working together and following the lead of the elder, teachers and students can create their patterns. The resulting pattern pieces can also be used as math manipulatives, since they include squares, rectangles, triangles, parallelograms, and rhombi. (Ilutsik and Zaslavsky expand on this strategy in the next article.) By turning students into producers, teachers are connecting the classroom with the everyday practice of the community. They are helping students learn mathematical relationships intuitively as the students practice fashioning their pattern pieces and learn the cultural rules governing this activity.

INFERRING CULTURAL RULES RELATED TO MATHEMATICS

By organizing curriculum around the Yup'ik seasonal subsistence cycle, including the skills and knowledge required to produce manipulatives, and incorporating the elders' expertise, we have begun fashioning a curriculum that is based on and in Yup'ik culture. A second approach that we have followed in our efforts to help students understand the indigenous mathematics is to examine and discuss the literal meaning of Yup'ik words. We have used this approach to understand and present Yup'ik numeration.

Numeration and Language

Our ethnographic and collaborative approach to working with Yup'ik elders has gone beyond giving elders the typical "show and tell" role. Instead, they assist in the construction of curriculum, actively working with teachers and university consultants in the process of translating their knowledge into classroom-based activities. Sometimes, as is true for enumeration, the aboriginal context may no longer exist. In such situations, we turn, together with groups of elders, to an examination of the Yup'ik language in an attempt to reconstruct the system. For example, working in Yup'ik, we use the literal meanings related to numeration to provide insights into Yup'ik numerical categories and conceptions. Here, the Yup'ik language supplies information about how numbers are organized and the implied cultural and mathematical rules associated with numeration. The examples from Yup'ik in table 24.1 provide the raw data from which we have developed culturally based mathematics activities.

As the table shows, the Yup'ik numeration system is a base-twenty system, with subgroups of five. Counting takes place within groups (one to five, five to ten, and so on) until the four groups of five are completed, forming a whole. The literal meaning of *yuinaq* (twenty) is "one person complete." We have gained insight into the construction of the Yup'ik number system not only through the literal meanings of the Yup'ik words associated with numeration but also through exposure to Yup'ik counting techniques. For example, we once asked elders to count a group of sticks. They arranged the sticks into groups of five and then bundled four groups of five into one bundle of twenty sticks.

Yup'ik elders count on the left hand first, moving from the pinky (or outside) finger to the thumb. As they count higher, they continue to make correspondences to parts of the body. Counting from one to five takes them from one finger to one arm—the literal meaning of *talliman*, the word used for five, is "arm." The Yup'ik word for six—*arvinlegen*—literally means "crossing over," implying in the counting context "crossing to the other hand." Ten, however, is not constructed in Yup'ik as the sum of two arms, although when the person counting reaches ten, he or she holds both hands at shoulder height near the body. In fact, *qula* (ten), or *qulen*, reflects the notion of reaching a midpoint—

Table 24.1
Yup'ik Cardinal Numbers and Literal Meanings

Counting on One Hand	Counting on the Other Hand
1 *atauciq*	6 *arvinlegen* (cross over)
2 *malruk*	7 *malrunlegen*
3 *pingayun*	8 *pingayunlegen*
4 *cetaman*	9 *qulngunritaraan* (not quite ten or not quite above)
5 *talliman* (one arm)	10 *qula, qulen* (above)

Counting Below on One Side	Counting Below on the Other Side
11 *qula atauciq*	16 *akimiaq atuaciq* (one on the other side)
12 *qula malruk*	17 *akimiaq malruk*
13 *qula pingayun*	18 *akimiaq pingayun*
14 *akmiarunritaraan*	19 *yuinaunritaraan* (not quite twenty or not quite a complete person)
15 *akimiaq* (the other side)	20 *yuinaq* (the whole person)

Counting beyond 20 shows how base twenty is incorporated in the language.

30 *yuinaq quala* (a whole person plus 'above,' or 20 +10)
40 *yuinaak malruk* (two wholes, or 20×2)

having traversed the top half of the body—as suggested in the literal meaning of the Yup'ik word *qula*—"above." The counting process continues on the lower half of the body: traditionally, eleven meant "going down to the feet on one side." A "whole person"—*yuniaq*—is reached at twenty. This description provides some insight into Yup'ik numeration and the mathematics encoded within the Yup'ik language and practice of counting (see Lipka [1998] for a more detailed discussion). Mathematics exercises that follow these literal meanings can make counting and adding small numbers very clear processes to speakers of Yup'ik, since the language shapes the mathematical activities. Thus, constructing such exercises is another approach we have been using in creating a mathematics curriculum.

Our way of working involves mainly "insiders"—Yup'ik elders and teachers—as well as a few "outsiders," such as academic consultants. Group members engage together in seasonal cultural activities, and members who are new to the customs learn first by observing and listening. Then, as they become ready, they participate, sometimes earning the designation "teacher" as they demonstrate competence. All group members perform the activities that the elders teach, such as making tools or weaving baskets. Instructions and demonstrations may be given in the local language, with translations for nonspeakers.

This approach allows learners to construct their own intuitive understandings and to develop competence through participation in expert-apprentice relationships. Lengthy discussions may ensue between learners and elders and among the elders themselves as the group strives for consensus on literal meanings and fundamental terms and concepts. Sometimes, as when local knowledge is fragmented, the group may call on ethnographic records as sources of information. Members' various modes of learning ensure that the group is gaining an authentic understanding of the knowledge that it is applying.

Our next step is to take these understandings and apply them to classroom situations. This stage gives full play to group members' creativity. We have used patterns to make headdresses used in dancing, created songs and dances using

CONNECTING CONTEXT AND CULTURE WITH THE CLASSROOM

a Yup'ik drum and traditional ways of teaching, and used the drum to beat out human heartbeats as a way of teaching numeration and patterning (see Ina White in Lipka [1998]). We have also used the drum as a tool in the teaching of grouping and numbers. Some of these activities involve the elders, who not only provide instruction through modeling and storytelling but also are part of the creative process of translating their knowledge into school settings.

AN INTERPRETIVE AND CREATIVE APPROACH TO CLASSROOM MATHEMATICS

In Yup'ik culture, the human body plays a central role as a cognitive construct and an instrument of sensation that facilitates counting, measuring, and judging. Spatial dimensions of the body are related to patterns (*top* and *bottom* and *left* and *right*), proportionality, and ratios. The following examples underscore the body's centrality.

A "Yup'ik" Abacus: An Interpretive and Creative Extension

Figure 24.2 shows an example of an ethnomathematical tool—an abacus—that has been designed to be appropriate for Yup'ik children in the lower elementary grades. As noted earlier, the abacus is not a part of traditional Yup'ik culture. However, this particular abacus is based on groupings embedded in the Yup'ik language and system of numeration. Its construction follows Yup'ik artistic and symbolic conventions, making the human form central and showing the circle and dot motifs that are quite common (Fienup-Riordan 1994). Thus, making the abacus can extend Yup'ik students' understanding of aspects of their culture and language. The groups of four beads followed by a stationary block on the abacus's outer circle parallel the Yup'ik pattern of counting within groups until a limit, such as one hand or the number five, is reached. At the same time, the abacus provides students with a first experience with a historically important mathematical tool, even if the word *abacus* is never applied to it. The educational tool acts as a bridge between culturally familiar ways of grouping and Western mathematical standards associated with grouping, counting, and adding. Here we have an example of the "both/and" approach in action: mathematics tools are set in a local context to forge connections to other cultural and mathematical contexts.

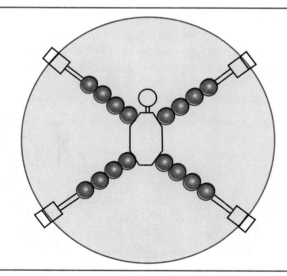

Fig. 24.2. Yup'ik Abacus ™

Patterns: Corresponding to, and Diverging from, the Culture

Patterns are another mathematical topic that we have pursued fruitfully with Yup'ik elders. In a number of meetings, the elders demonstrated how they produce patterns for such uses as borders on their clothing. Central to their approach to pattern making is the establishment of an initial square. One way in which they form this square is by overlapping two pieces of fabric or fur (see fig. 24.3). The resulting square then becomes the shape from which all other shapes are formed, and all the shapes are then related to one another. Figures 24.3 and 24.4, which involve pattern making, help illustrate our method of developing ethnomathematics activities. These activities enable students to learn about, construct, manipulate, and expand typical Yup'ik border patterns.

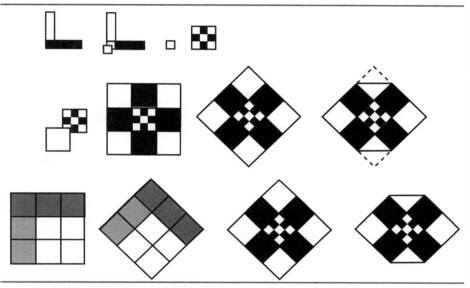

Fig. 24.3. One way to create border patterns

In the process shown in figure 24.3, the choice of the geometrical patterns and the way in which they are produced draw on knowledge of traditional Yup'ik culture and set the activity in a Yup'ik context. Our thematic and contextualized mathematics module for the classroom uses the Yup'ik names for these patterns (as Ilutsik and Zaslavsky describe in detail in the following article). Students create and apply these patterns in traditional ways. Moreover, visualizing and estimating body measures and observing elders as they construct traditional patterns are important parts of the module. By incorporating the elders' methods, the mathematics lesson acquires the power to transmit Yup'ik culture and language.

Fig. 24.4. Plus-sign pattern

Figure 24.4 shows how these same geometrical shapes can serve as pattern blocks that can be recombined to create shapes not typically found in Yup'ik culture. Students learn intuitively about the relationship between the "generator" square and the rest of the design (fractals). By examining and making other traditional designs that include triangles and rectangles, students learn experimentally about their relationships, as well. We encourage teachers and students to create their own Yup'ik pattern blocks. Doing so involves students in the production process—planning, designing, and creating. Their experiences strengthen their growing sense of geometric relationships. The pattern blocks can be manipulated like any other pattern-block set. However, the traditional Yup'ik patterns have symbolic and cultural meaning that add another dimension to this ethnomathematic tool. (Ilutsik and Zaslavsky explore this topic in more detail in the next article, and Lipka [1998] describes other uses of the pattern blocks.)

Second-Order Tools: Moving Away from Yup'ik Culture

Working from these Yup'ik patterns provides an opportunity to extend the mathematical power of this tool but with a weakened connection with Yup'ik culture. Figure 24.5 shows some of the patterns that can be made from the "plus sign" pattern shown in figure 24.4. The plus-sign pattern can be visualized as containing these designs, which we can describe as diagonals, "straights," L-shapes, steps, and V-shapes. Except for the plus sign itself, these patterns are not among those that we have seen elders construct. All the other patterns are simply possible. They are directly connected to the Yup'ik plus-sign pattern, yet they diverge from traditional Yup'ik patterns and conceptions. The task of finding other patterns contained within the plus-sign design is an exercise in visualizing and creating new sets of mathematics tools. Reconstructing the pattern with these new pieces presents fresh challenges to the students. Also, these pieces can be used by students in designing Yup'ik and non-Yup'ik patterns.

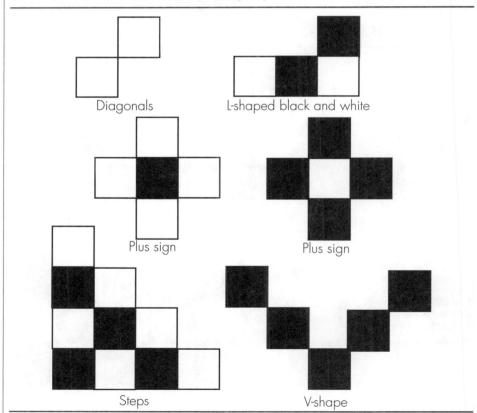

Figure 24.5. Possible patterns

The new tools—the new pattern blocks—form a bridge from the Yup'ik culture to the larger culture and its conventions and rules. Having students make such tools is an example of what we have been calling a second-order ethnomathematics activity. Our process of developing the activity illustrates our method of producing an ethnomathematics curriculum that is conceptually grounded within Yup'ik culture and reinforces that culture while simultaneously allowing students to grow academically in other directions.

CONCLUSIONS

Our ethnographic and collaborative approach to developing an ethnomathematics curriculum is based in Yup'ik culture and mathematics and shows how indigenous teachers and their university colleagues can learn from Yup'ik elders through participatory research. This approach to curriculum and pedagogy simultaneously reinforces local culture and language while building an intuitive understanding of measuring, geometry, and numeration. Thematic modules related to seasonal activities connect our mathematics curriculum with Yup'ik culture. By incorporating work that elders do as producers, our classroom curriculum enables students to learn some of their skills and knowledge. These relate directly to geometry, visualizing, estimating, and proportionality. By involving the teachers and elders in adapting traditional knowledge to schooling, our process becomes emancipatory, enabling teachers and students to appreciate the wisdom of the elders. Gerdes (1997, p. 239) suggests that

> [b]y this process of rediscovering the mathematical thinking hidden in these baskets and fishtraps—and in other traditional production techniques—the future teachers feel themselves stimulated to reconsider the value of their cultural heritage: in fact, geometrical thinking was not and is not alien to their culture.

By grounding knowledge in the local tradition and adapting, interpreting, and creating classroom activities that retain vital connections with the culture, the context, and the indigenous language, ethnomathematics can serve as a bridge between a local indigenous culture and the larger society.

Although our approach to formulating curriculum for classroom use does not resolve all the curricular dilemmas outlined at the beginning of this article, it does address a substantial number. What is most important is that the approach begins and ends with the local indigenous people. Our development of an ethnomathematical curriculum is grounded in the culture through our work with elders. This approach recognizes the still-breathing, -evolving, and -changing context of Alaska Native cultures and incorporates it into the design of curricula. Elders and their ways serve as a foundation for the curriculum and its tools, yet these do more than hold up a mirror to that culture. This type of curriculum offers schools and communities new choices. The next article explores these opportunities and ideas more fully at the level of classroom instruction.

REFERENCES

Bazin, Maurice, and Modesto Tamez. *Math across Cultures*. San Francisco: The Exploratorium, 1995.

Bradley, Claudette. "Issues in Mathematics Education for Native Americans and Directions for Research." *Journal for Research in Mathematics Education* 15 (March 1984): 96–106.

———. "Making a Navajo Blanket Design with Logo." *Arithmetic Teacher* 40 (May 1993): 520–23.

Bruner, Jerome. *The Culture of Education.* Cambridge, Mass.: Harvard University Press, 1996.

Caduto, Michael, and Joseph Bruhac. *Keepers of the Earth.* Golden, Colo.: Fulcrum, 1988.

Denny, J. Peter. "Cultural Ecology of Mathematics: Ojibway and Inuit Hunters." In *Native American Mathematics,* edited by Michael Closs, pp. 129–80. Austin, Tex.: University of Texas Press, 1986.

Fienup-Riordan, Ann. *Boundaries and Passages: Rule and Ritual in Yup'ik Eskimo Oral Tradition.* Norman, Okla.: University of Oklahoma Press, 1994.

Gay, John, and Michael Cole. *The New Mathematics and an Old Culture: A Study of Learning among the Kpelle of Liberia.* New York: Holt, Rinehart, & Winston, 1967.

Gerdes, Paulus. "On Culture, Geometrical Thinking, and Mathematics Education." In *Ethnomathematics,* edited by Arthur Powell and Marilyn Frankenstein, pp. 223–47. Albany, N.Y.: State University of New York Press, 1997.

Gettman, David. *Basic Montessori: Learning Activities for Under-Fives.* New York: St. Martin's Press, 1987.

Hollins, Etta. *Culture in School Learning: Revealing the Deep Meaning.* Mahwah, N.J.: Lawrence Erlbaum Associates, 1996.

Kawagley, Oscar. *A Yupiaq Worldview: A Pathway to an Ecology and Spirit.* Prospect Heights, Ill.: Waveland Press, 1995.

Lipka, Jerry. "Culturally Negotiated Schooling: Towards a Yup'ik Mathematics." *Journal of American Indian Education* 33, no. 3 (1994): 14–30.

———. *Transforming the Culture of Schools: Yup'ik Eskimo Examples.* Mahwah, N.J.: Lawrence Erlbaum Associates, 1998.

Lipka, Jerry, and Teresa McCarty. "Changing the Culture of Schooling: Navajo and Yup'ik Cases." *Anthropology and Education Quarterly* 25, no. 3 (1994): 266–84.

MacPherson, Jennifer. "Norman." *For the Learning of Mathematics* 7, no. 2 (1987): 24–26.

Meade, Marie. "Sewing to Maintain the Past, Present, and Future." *Inuit Studies* 14, no. 2 (1990): 229–39.

Pinxten, Rik. "Applications in the Teaching of Mathematics." In *Ethnomathematics,* edited by Arthur Powell and Marilyn Frankenstein, pp. 373–401. Albany, N.Y.: State University of New York Press, 1997.

Pinxten, Rik, Ingrid van Dooren, and Frank Harvey. *Anthropology of Space: Explorations into the Natural Philosophy of the Navajo.* Philadelphia: University of Pennsylvania Press, 1983.

Scollon, Ron. "Bush Consciousness and Modernization." In *Linguistic Convergence: An Ethnography of Speaking at Fort Chipewyan, Alberta,* by Ron Scollon and Suzanne Scollon, pp. 177–209. New York: Academic Press, 1979.

Zaslavsky, Claudia. *The Multicultural Math Classroom: Bringing In the World.* Portsmouth, N.H.: Heinemann, 1996.

Yup'ik Border Patterns in the Curriculum

25

Esther A. Ilutsik
Claudia Zaslavsky

Two of the most exciting developments in education today are the recognition of the culture of the community and the integration of information about that culture into the local school curriculum (see the article by Lipka, chapter 24 in this volume). This article deals with the mathematical aspects of the patterns displayed on traditional Yup'ik Eskimo fancy squirrel parkas. Much of the information is based on the wisdom shared by the elders of the Yup'ik community.

FANCY SQUIRREL PARKAS

Among the Yup'ik Eskimos, many different styles of parkas provide warmth against the winter cold of southwest Alaska. One is the fancy squirrel parka, which has decorations and patterns that identify the family of the wearer and are recognizable by members of the extended family and other villagers when the wearer travels to another village, perhaps for a winter celebration. Such recognition ensures proper treatment according to the strict kinship or community social system. Not only does the fancy squirrel parka reveal the person's identity, but it also served in the past to determine the worthiness of a young woman of marriageable age!

The parka (see fig. 25.1) is made from ground squirrel skins—about sixty skins for an adult male's parka, fifty for an adult female's parka, and forty or fewer for a child's garment. The decorative tassels and ruff are made from wolf or wolverine, but beaver is used for the bottom and cuff trimmings. Today, cowhide or calfskin is used for the border patterns. Squares, triangles, and other shapes are cut from black and white skin of caribou and seals and combined to form the borders. Beads and red yarn may be added to the decorations.

The types of animal skins are chosen for their distinctive properties. Squirrel skins are warm but lightweight. Wolf and wolverine protect the face from extreme cold; although these furs collect moisture, they tend to release water when the temperature drops below the freezing point, thus preventing a buildup of cold moisture. The heavy weight of the beaver trim helps keep the parka in place in the face of gusts of cold wind.

The patterns displayed on the border and the middle main tassel reveal either the maternal or paternal (depending on geographic location) identity of the wearer. The middle main tassel shows the symbol of the extended family—a beluga whale tail, a Yup'ik drum, the middle finger, and a river are some examples. Among the border patterns indicating the wearer's ancestry (see fig. 25.2) are *egaleruaq* (pretend window), *ingriruaq* (pretend mountain), *yaassiiguaq* (pretend box), and *taquruaq* (pretend braid). The ending *-uaq* of each noun indicates that it refers to a "pretend" rather than an actual object. Another common symbol is *kuiguaq* (pretend river), often associated with the pretend mountains. Found on many of the parkas are the *ikamram tumeliri* (sled tracks), a very important aspect of Yup'ik culture. The patterns discussed in this article are just a few of the many patterns displayed on fancy squirrel parkas.

Fig. 25.1. A fancy squirrel parka

ingtituaq (pretend mountain)

yaassiiguaq (pretend box)

ikamram tumellri (pretend sled tracks)

egaleruaq (pretend window)

kelistaruaq (pretend cross)

kevirun (filler)

ikamram tumellri (pretend sled tracks)

ingriruak (two mountains) or taksurenqellria (wanting to be tall)

ikamram tumellri (pretend sled tracks)

taquruaq (pretend braid)

ingtituaq (pretend mountain)

kuiguaq (pretend river)

ikamram tumellri (pretend sled tracks)

ingtituaq (pretend mountain)

ingriruak (two mountains) or taksurenqellria (wanting to be tall)

ikamram tumellri (pretend sled tracks)

cen'a (shore, border)

Fig. 25.2. *Tumaqcat* or *kepuqcat* (Yup'ik border pattens). The symbols in the center were documented by Mary George and Marie Napoka. The Ciulistet group and Mary George have adapted these traditional shaped for mathematics instruction. The Yup'ik terms include the word for "pretend" as a suffix to differentiate between real objects and symbols of them.

INTRODUCING BORDER PATTERNS INTO THE PRIMARY SCHOOL CURRICULUM

Two important questions arise: What areas of the school curriculum should the information about Yup'ik border patterns address? How can this material be adapted to fit Western-oriented schooling in a format that retains the traditional Yup'ik philosophy of teaching?

The topic is introduced at the kindergarten and first-grade levels in the Dillingham (Alaska) City Schools. The Yup'ik border patterns include the basic polygonal shapes that are familiar to children—the triangle, the square, the nonsquare rectangle, and other parallelograms. The square, the nonsquare rectangle, and the triangle are the shapes most often introduced at the primary school level, whereas the parallelogram is taught at a later stage.

A poster displaying the Yup'ik border patterns becomes part of the classroom decor. The patterns, each with its appropriate Yup'ik name, are depicted in black and white against a red background. Nearby is another poster showing adults and children wearing parkas with these patterns. In one of the classroom learning centers is a set of cards and a small poster with information about the border patterns, the geometric shapes, and the names of the patterns. A small paper doll in the center demonstrates some of the border patterns that can be made on the doll's paper parka. The children are asked to make the shape displayed on one of the cards.

The main poster appears in a prominent place in the classroom a week or two prior to the setting up of the center, thus addressing a traditional Yup'ik method of learning—observation. Children notice it as they line up to go to recess, to lunch, and to assembly. They begin to question the teacher and often recognize some of the patterns. They make such comments as "Oh, my mom made a parka with that pattern on it. What is it called?" or "I saw that pattern at the museum." In this way the teacher introduces the children to the patterns in a non-threatening situation. She may focus on one pattern, emphasize its Yup'ik name, and make a game of identifying it. She might ask, "Who can remember the name of the pattern Kris shared with us yesterday?" A student who remembers has the opportunity to choose another pattern, and the game continues the following day.

After about a week of this type of activity, the teacher sets up the centers, which may remain for as long as a month. The students visit the center once a week for fifteen to twenty-five minutes. In this way, the teacher again employs the traditional Yup'ik style of learning, allowing each child to learn at his or her individual pace after the observation stage. Assessment is also done individually, as the teacher observes the students' proficiency in reproducing specific border patterns and in identifying them by name.

INTEGRATING MATHEMATICS AND THE CULTURE

In second grade, more-formal Western pedagogy is incorporated into the unit. The children review the basic shapes found in the Yup'ik border patterns, starting, as before, by viewing appropriate posters displayed in a high-use area of the classroom. A discussion of the basic shapes becomes an integral part of the mathematics lessons.

The students develop their oral and listening skills as they play the "back-to-back" game: Two children sit on the floor back-to-back, each with a set of pattern blocks or other precut shapes. Out of the view of her partner, one child creates a pattern using these shapes. When she has completed it, she gives verbal directions to the other child to reproduce the pattern she has created. The children then compare their shapes. They discuss whether discrepancies are due to an inadequate verbal description of the pattern or to the failure of the second child to follow directions. The children discuss how they might improve the procedure. Teachers introduce the game by having two children give a demonstration to the others. When they understand, the game continues in the learning center for further practice.

Following this introduction, the class moves on to discuss the patterns they find in the real world. Of greatest interest are the patterns that are distinguishing features of various animals. Children in Alaska are familiar with the ptarmigan. What pattern do they see on the bird's tail? A child might find that it has fifteen feathers that are gray-black with brown tips and fifteen that are black with white tips. What is the sum of the two groups of feathers? Do all ptarmigans have the same pattern? The question Why or why not? may well lead to a science research project.

After analyzing the patterns in the children's clothing, in classroom objects, and in the American flag, the teacher introduces the role of patterns in the lives of the Yup'ik ancestors. The discussion is focused on the fancy squirrel parka—how it is made and the significance of the various parts. The teacher selects one specific border, say, the *egaleruaq* (pretend window). The children examine the poster depicting this border and discuss the components, their shapes and colors, and the number of each. What do they think the pattern means? It is called *egaleruaq*—a pretend window—because many years ago the Yup'ik lived in sod houses with a single window. It was located in the center of the home, in the roof, depicted by the white square in the center of the pattern, and supported by four beams, the black squares (fig. 25.3).

Fig. 25.3. The *egaleruaq* (pretend window pattern)

After the children have discussed the number of squares of each color and the total number required to make one *egaleruaq*, one unit of the pattern, they are given the appropriate number of black and white squares and sheets of red construction paper with which they form their own versions of the pretend window. Later they may be asked to compute the number of black squares and of white squares necessary to form two *egaleruak* (*-uak* is the ending of the word that indicates *two*) and three or more *egaleruat* (*-uat* is the plural ending of the word).

One of the most inspiring aspects of this integration of the local culture into the curriculum is the enthusiasm of the children. Here are several typical remarks:

- "I never knew math could be so much fun."

- "My mom wants to bring in her Yup'ik parka; she has a pattern on it that she wants to know more about."

- "Can I bring my mother to school so that she can see what we are doing?"

The teachers are often impressed by the students' performance and the change in their attitude toward school and themselves. Students who had previously done poorly often show marked improvement, as illustrated by one teacher's remark, "Moses has been having a hard time in school, but this really turned him on to learning. Maybe it is because this is about his culture and he is more accepting about who he is."

After the units, children become aware of shapes and patterns outside the school boundaries and come to class enthusiastic about sharing their newly acquired information. One student related that as she was waiting in the doctor's office, she recognized the *egaleruaq* pattern on the curtains. Students with Yup'ik ancestry have a better understanding of who they are and are motivated to seek more information about themselves and about Yup'ik culture and language. Those with other cultural backgrounds have a new respect for the Yup'ik culture and become more curious about their own heritage.

Yup'ik elders play an integral role in curriculum planning. At a three-day conference arranged by University of Alaska instructors for teachers and teacher's aides in predominantly Yup'ik schools, and for Yup'ik elders, one afternoon was devoted to a discussion of border patterns and symmetry (Zaslavsky 1996, pp. 149–52). All the participants were women; the men were in another room discussing the Yup'ik calendar. Each participant received a packet of precut posterboard squares and isosceles right triangles, black and white, in two different sizes. One teacher had brought in a set of squares and triangles made of black and white calfskin, along with needles, thread, and scissors.

After a discussion of symmetry and similarity, led by a university instructor, the group worked in twos to create border patterns, some traditional, some more original. Then each pair shared with another pair an analysis of the creation from both the mathematical and cultural points of view. What types of symmetry were involved in repeating the patterns to extend them all along the bottom of the parka? Similarity was illustrated with reference to design units, such as a comparison of *egaleruaq* (pretend window) with *yaassiiguaq* (pretend box) (fig. 25.4). Note that a change in the size or orientation of the design elements is associated with a different interpretation.

It was a proud moment for one of the elders when her daughter, a bilingual teacher, displayed two parkas that the mother had made many years before. These garments became the focus of attention as the group applied their mathematical knowledge to an analysis of the patterns.

A lively discussion ensued in both Yup'ik and English to determine the proper Yup'ik terminology for the mathematical words used to describe symmetry and similarity. The Yup'ik elders, some of whom did not speak English, were consulted, and the group finally came up with the following:

Symmetry: *cuq yut kutatekellriit* (same measure)

Rotation: *uivurallra*

Translation: *ayuguralriit*

Reflection: *ayaqiin akia* (a term used in sewing) or *tarenraa* (mirror)

Shrink, stretch: *angligtaalra, mikligtaallra*

One bilingual teacher remarked that in her region the vocabulary was slightly different, and she would have to consult the elders of her community for the proper terminology.

The teachers and teacher's aides set to work to write lesson plans for their specific grades, basing the lesson on the use of manipulatives and incorporating the mathematical and cultural concepts as well as the Yup'ik terminology that had been discussed. Each teacher or teacher's aide explained her lesson, using the overhead projector to display the patterns. The session concluded with the introduction of the "back to back" game described earlier.

A consultant to the group showed the connection between their work and ethnomathematical research in Africa by passing around the book *Sipatsi: Technology, Art, and Geometry in Inhambane* (Gerdes and Bulafo 1994). It deals with certain types of woven handbags (*sipatsi*) produced in a province of Mozambique and now attracting national and international attention. The book reproduces many of the traditional designs and repeated patterns found on the *sipatsi* and is intended as a reference for a course called "Geometry and Symmetry." The Yup'ik women, especially the elders, were visibly thrilled to learn about the recognition given traditional work and recognized a

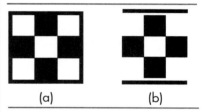

(a)　　　　(b)

Fig. 25.4. Design elements for (a) the *egaleruaq* (pretend window) and (b) the *yaassiiguaq* (pretend box)

resemblance to the patterns in their own basketry. They would probably have been even more excited about a later book, *Women and Geometry in Southern Africa* (Gerdes 1995). One chapter, "Pythagoras a Woman? Example of an Educational-Mathematical Examination," deals with proofs of the right-triangle theorem ($a^2 + b^2 = c^2$; see the next section of this article) based on a pattern called the "toothed square" in the coiled basketry of the Ovimbundu women of Angola.

THE YUP'IK WAY IN THE MATHEMATICS CURRICULUM

We live in a consumer society. Contrast the school's precut shapes and pattern blocks with the Yup'ik practice of cutting all the necessary components from raw materials. A Yup'ik woman works out a unit square for each application (fig. 25.5a). To form the parallelograms for the *taquruaq* (pretend braid) pattern, she starts with a square and cuts along lines slanted at the appropriate angle (fig. 25.5b). The planning that goes into the production of these shapes and their placement along the border is a mathematically valuable aspect of pattern making. When children engage in such activities, the mathematical concepts they are expected to learn become obvious to them. Cutting the shapes and placing them on the border requires careful measurement of lengths and angles. Naturally, two congruent isosceles right triangles can be joined to form a square because originally they were cut from a square. Such concepts as area and perimeter, proportionality, congruence, and similarity become clear in the course of working with the materials.

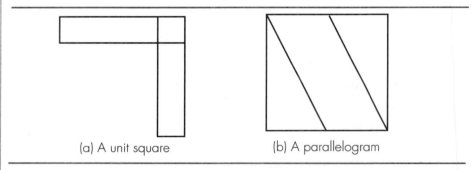

(a) A unit square   (b) A parallelogram

Fig. 25.5. Elements of Yup'ik border patterns

To go back to "Pythagoras a Woman?" how can we use the shapes to prove that in a right triangle, the sum of the squares on the legs is equal to the square on the hypotenuse? Let us start with the special case of an isosceles right triangle in which each leg has a length of n units.

- Form a $2n \times 2n$ square of black unit squares, or use a square cut to these dimensions.
- Cut two $n \times n$ white squares along a diagonal to form four isosceles right triangles. Place them on the black square so that they form an inscribed square (fig. 25.6).
- In triangle ABC, leg AC and leg BC each form one side of a square with area n^2. The sum of the areas of the two squares is $2n^2$.
- In triangle ABC, the hypotenuse AB forms one side of a square with area $2n^2$ (the combined areas of the four white triangles).
- Therefore $BC^2 + AC^2 = AB^2$, or $a^2 + b^2 = c^2$.

It may be simpler to start with unit squares, that is, to let $n = 1$ in the proof above.

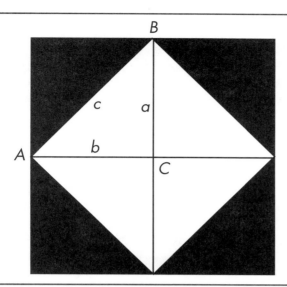

Fig. 25.6. An argument for the right-triangle theorem based on Yup'ik culture

The "inscribed square" proof of the right-triangle theorem is just one example of the possibilities for developing a Western style of mathematics curriculum based on the Yup'ik way of dealing with mathematical concepts.

CONCLUSION

The National Council of Teachers of Mathematics's *Curriculum and Evaluation Standards for School Mathematics* (1989) recommends greater emphasis on geometry, measurement, and pattern recognition in the early grades. Relating to the value of geometric thinking, the document states, "Children who develop a strong sense of spatial relationships and who master the concepts and language of geometry are better prepared to learn number and measurement ideas" (p. 48). Curriculum materials based on repeated patterns in Yup'ik parka borders fulfill the recommendation admirably. Students handle physical materials, observe the relationship of one shape to another and to the entire pattern, and they communicate their ideas about mathematics to their fellow students.

At the same time, they learn to appreciate the cultural values inherent in the patterns. People in all parts of the world and in all eras of history have developed mathematical ideas because they needed to solve vital problems in their lives. Children learn that mathematics plays an important role in society, both in their own and in other people's cultures.

REFERENCES

Gerdes, Paulus. *Women and Geometry in Southern Africa.* Maputo, Mozambique: Universidade Padagogica Mocambique, 1995.

Gerdes, Paulus, and Gildo Bulafo. *Sipatsi: Technology, Art, and Geometry in Inhambane.* Maputo, Mozambique: Instituto Superior Pedagogico Mocambique, 1994.

National Council of Teachers of Mathematics (NCTM). *Curriculum and Evaluation Standards for School Mathematics.* Reston, Va.: NCTM, 1989.

Zaslavsky, Claudia. *The Multicultural Math Classroom: Bringing In the World.* Portsmouth, N.H.: Heinemann, 1996.